Leaf Defence

EDWARD E. FARMER

Great Clarendon Street, Oxford, OX2 6DP,
United Kingdom

Oxford University Press is a department of the University of Oxford.
It furthers the University's objective of excellence in research, scholarship,
and education by publishing worldwide. Oxford is a registered trade mark of
Oxford University Press in the UK and in certain other countries

Impression: 1

Published in the United States of America by Oxford University Press
198 Madison Avenue, New York, NY 10016, United States of America

British Library Cataloguing in Publication Data
Data available

Library of Congress Control Number: 2013957441

ISBN 978-0-19-967144-1

Printed in Great Britain by
Clays Ltd, St Ives plc

Preface

Leaves are highly abundant and important biological units and are the foundation of the terrestrial limb of the global carbon cycle. But their principal function to capture and reduce carbon, coupled to their often delicate structures, makes them vulnerable to predation. Why, then, wherever there is enough water for plant growth, are leaves are so abundant? In other words, how do leaves defend themselves from myriads of herbivores? This book is meant to be a broad, non-exhaustive introduction to leaf defences against leaf-eating vertebrates and invertebrates, and is written for anyone with an undergraduate training in biology. It has been my intention to avoid producing a reference work in which all the possible defence mechanisms used by leaves are listed. Moreover, with one or two exceptions, the defences of roots, flowers, seeds, and fruits are not discussed. The idea is to concentrate on leaf defences and to do this by combining two complementary outlooks: one from my professional life as an experimental molecular biologist, the other reflecting an interest in observational field biology. The writing reflects this. In a few places a more formal text gives way to anecdotes that describe leaves or leaf eaters in various parts of the globe; these are recounted in the first person. Throughout, I have tried to use an evolutionary approach without which much of nature's great beauty, both in the field and in the laboratory, is lost on me.

The book is organized so that it moves from macroscopic aspects of leaf defences towards the molecular level. That is, after a general introduction on leaves, herbivores and the extent of herbivory, the text then transitions from the impact of plant growth and form on defence through to the plant's use of specialized defence cells and, finally, to the molecular mechanisms of defence. Then, after a section on wound-inducible leaf defence mechanisms I turn to some of the partnerships between plants and the organisms that protect them. Towards the end of the book I examine anachronistic plant defences—those defence traits that show us how plants defended themselves from leaf eaters that no longer exist. Anachronistic defences have been studied in depth on several island groups, most notably on New Zealand and Hawaii. However, for some of the botanically rich islands in the Indian Ocean the story is different and we do not know whether plant-eating animals existed on certain islands prior to human arrival. In the search for these biological ghosts I use the last main chapter in an attempt to interpret putative anachronistic defences on Socotra, an island off the coast of Yemen. In making this journey to and from the laboratory and the field I have tried to place emphasis on botany, which for me is as much an appreciation of individual plant species as it is of plant families. So, as much as is possible, I mention families of plants. In most cases this is not intended to mean that the family is unique, but there are some families that seem to be more creative in their use of defences than others. Inevitably, those that I particularly like get more than their fair share of attention.

A difficulty related to this is that the field of taxonomy is, at present, moving fast. The names of genera change quickly and some family names disappear altogether. An effort has been made to keep things up to date, although this can sometimes be awkward. This is, for example, the case when it comes to acacias, many of which have been reclassified from *Acacia* to *Vachellia*. I try to strike a balance between the new and the old here, often keeping acacia as the vernacular name (particularly when African plants are discussed).

At the outset I'd like to acknowledge several books that have been both useful and inspirational: M.J. Crawley's *Herbivory: The Dynamics of Animal-Plant Interactions*, (Blackwell, 1983); J.B. Harbourne's *Introduction to Ecological Biochemistry*, 4th edition (Academic Press, 1993); Richard Karban and Ian T. Baldwin's *Induced Responses to Herbivory* (University of Chicago Press, 1997); *Induced Resistance to Herbivory*, edited by Andreas Schaller (Springer, 2008); Schoonoven, van Loon, and Dicke's *Insect-Plant Biology*, 2nd edition (Oxford University Press, 2005); and Dale Walters' *Plant Defence: Warding off Attack by Pathogens, Herbivores, and Parasitic Plants* (Blackwell, 2011). This later book contains an up-to-date synthesis of the various theories of plant defence, a subject only briefly touched on herein. For field trips I have used and been inspired by several field guides including Jonathan Kingdon's *The Kingdon Field Guide to African Mammals* (Academic Press, 2001), R.D. Estes' *The Behavior Guide to African Mammals: Including Hoofed Mammals, Carnivores, Primates* (University of California Press, 1992), and John Kricher's *A Neotropical Companion* (Princeton University Press, 1997).

Now to the people who have helped me gain experience in biology. I owe, first and foremost, a great deal to the late Clarence ('Bud') Ryan to whom I was drawn in order to enter the fascinating field of plant-herbivore interaction and to search for regulatory molecules underlying inducible defence responses. Another person who both influenced and encouraged me was the late Meinhard Zenk. With Bud Ryan's support I was able to glimpse my first tropical forests in Brazil and Costa Rica, and Meinhard Zenk paved the way for an extended visit to New Caledonia where I was generously accommodated at the Institute de Recherche pour la Dévéloppement (IRD) in Noumea. For this, for some fascinating glimpses of the New Caledonian flora, and for much discussion, I'm particularly grateful to Tanguy Jaffré and Maurice Schmid. Vincent Dumontet organized a memorable field trip and I thank Yves Létocart and the staff of the 'Parc de la Rivière Bleue' for their generous help. I benefited from the company of Douglas Tallamy in Brasil and E. Durant McArthur in Utah and I owe a great deal to Brian Barlow, William (Bill) Foley and Andrew Krockenberger for nurturing my interest in the Australian flora and fauna. In New Zealand I thank Ian Payton, Mat McGlone, and David Glenny for a series of rewarding field trips, William Russell (Bill) Sykes for stimulating discussions, and Mary Korver for generous help in identifying and shipping specimens. Concerning Socotra I'm indebted to Anthony (Tony) G. Miller and Lisa Banfield, both from the Royal Botanic Garden Edinburgh (RGBE), for their help and plant identifications and for much useful discussion, to Ahmed Saeed Suleiman for his assistance in the field, and to Kay van Damme for useful advice. Follow-up work at RGBE for the Socotra trip was financed by a European Union Synthesis grant and facilitated by Alan Forrest, Sabina Knees, and John Mitchell. My thanks also go to Khaled Al-Rasheid, Ahmed H. Al Farhan, and their colleagues at King Saud University in Saudi Arabia for their support in organizing a field trip to southwestern Saudi Arabia and the Farasan Islands. Tim Blackburn, also on this trip, kindly shared his insights into bird-plant interactions. Jacob Thomas at the KSU University herbarium in Riyadh provided expert help and good company while in the field and has since answered many questions concerning the Saudi flora.

In Gabon I benefited greatly from discussions with John Terborgh, Lisa Davenport, Raoul Niangadouma, and Patrice Christy. Related to this trip, I thank Melanie L.J. Stiassny of the American Museum of Natural History for fish identification. I should also mention my excellent pirogue team of Maxi, Jacques, and Lucien. I thank Raphaël Jordan for guidance in Ethiopia. Back in Europe Dénis Jordan and Gilles Yoccoz kindly shared unpublished information on the flora and fauna of the *Jardin du Talèfre*, and Pascal Vittoz is thanked both for his company at the *Jardin* and elsewhere in the Alps, as well as for his insightful thoughts on plant survival. Philippe Danton generously shared his expertise on France's Mediterranean flora and brought some remarkable books and plants to my attention.

For certain more molecular aspects I have drawn on my own laboratory's research which has been supported by the Swiss National Science Foundation, The University of Lausanne, King Saud University (Riyadh), and the Leenaards Foundation. I also gratefully acknowledge support from the European Molecular Biology Organization and the European Union for financing postdoctoral fellowships in my laboratory. The section on jasmonate signalling and the origins of herbivory is condensed from the 2009 Loomis lecture that I gave at Iowa State University. Some of the other content is from lectures given at the University of Lausanne and elsewhere. I thank the members of my own laboratory at the University of Lausanne, past and present. In particular, I'm grateful to Johann ('Hans') Weber and to Philippe Reymond for their help running the laboratory while I was consumed with the task of being departmental director. It is a great pleasure to acknowledge the support of my long-serving and excellent technicians Aurore Chételat and Stéphanie Stolz. Others in Lausanne have provided useful insights. Among these people are Lionel Maumary, Alia Malfi, Roland Keller, Daniel Cherix, Laurent Keller, and Philippe Christe.

In addition to the people mentioned above, numerous colleagues over the years have taught me about plant defence. These people include Anurag Agrawal, Ian Baldwin, Carlos Ballaré, Wilhelm Boland, Regine Claßen-Bockhoff, Rod Croteau, Marcel Dicke, the late Thomas Eisner, Matthias Erb, Gary Felton, Jonathan Gershenzon, David Giron, Ikuko Hara-Nishimura, Martin Heil, Daniel H. Janzen, Thomas Juenger, Richard Karban, Richard Lindroth, Consuelo de Moraes, Corné M.J. Pieterse, Bill Proebsting, Sergio Rasmann, Jennifer Thaler, Jim Tumlinson, Ted Turlings, Linda Walling, and close colleagues from the jasmonate field including John Browse, Mats Ellerström, Alain Goossens, Gregg Howe, Kemal Kazan, Martin G. Mueller, Hiroyuki Ohta, Roberto Solano and Jean-Luc Wolfender. Bernd Nilius introduced me to the fascinating world of transient receptor channels. When it came to putting the book together, Antonio Muccioli, Bruno Humbel, and colleagues at the Lausanne Electron Microscope Facility produced many useful images. Thomas Degen is thanked for his careful artwork, Geoffroy Colau for aid in preparing figures, Isabelle Tschou for help with references, and Josiane Bonetti for the skilled retrieval of out-of-print works. I thank the OUP team and in particular Lucy Nash, Robin Lewis Watson, and Smita Gupta for their support and patience. Finally, I am grateful to Mieke de Wit, Stephan Kellenberger, Pascal Vittoz, and John Pannell for valuable criticism of parts of the manuscript. Inevitably, trying to bring both laboratory results and field observations together results in some sections being superficial and, of course, each and every mistake and misinterpretation is my own. For this I apologize, as I do to colleagues whose work was not cited.

Edward E. Farmer
Lausanne, Switzerland

Contents

1. Introduction: the leaf and the pressures it faces 1
Herbivory and the terrestrial carbon cycle 2
The range of leaf-eating organisms to be considered 4
Herbivores outnumber plants 5
Vertebrate folivores 11
The nature of the leaf 17
Summary 26

2. Leaf colour patterning and leaf form 29
The palette of colour available to leaves 29
Folivore exposure colouration 31
Non-deceptive colouration: warning 33
Deceptive colouration: colonization and damage mimicry 37
Dissimulation and crypsis 40
Crypsis through movement 44

3. Structural defences and specialized defence cells 47
Leaf surface defence cells and leaf surface habitats 47
Stings 52
Glochids, hairs, prickles, and thorns 55
Silica: targeting teeth and mandibles 62
Crystalline defences 65
Fibre cells and sclereids 68
Defence cells: idioblasts 70
Exudates 70
Summary 73

4. Chemical defences 75
Selection for novel defence chemistries 75
Alkaloids 86
Phenolics 92
Detoxification: the important tail end of the defence process 100
Terpenes 104

5. Inducible defences and the jasmonate pathway 115
Inducible proteins that deplete energy and essential nutrients 115
The moving defence horizon 119
Activating inducible defence: the importance of having good teeth 120
The jasmonate pathway 122
Jasmonate and growth 132

The evolution of jasmonic acid-based signalling 135
The suppression of jasmonate signalling by herbivores 138

6. Top-down pressures and indirect defences 141
Plant population remodelling after carnivore removal 141
Ant–plant interactions, extrafloral nectaries, and food bodies 143
Mite domatia 147
Predator and parasitoid attraction by plant volatiles 150
Kudus and acacias: a cautionary tale 154

7. Release and escape from herbivory 155
Release from herbivory: anachronistic defences 155
Escape from vertebrate folivores 161

8. Escape in space: the cliff trees of Socotra 165
The cliff trees of Socotra 165
Candidate invertebrate herbivores on Socotra 172
Candidate vertebrate herbivores on Socotra 172

Synthesis 177
Wealth and social context 177
Tolerance and escape 177
Defence expenditure is proportional to attack pressure 178
The time factor in defence 178
Energetics: leaves and digestive tracts have co-evolved 178
Defence inducibility and the jasmonate pathway 179
There is no single best defence strategy 179
Leaves co-operate with the enemies of their enemies 180
Relevance of plant defences in agriculture and industry 180

Glossary 183
References 187
Index 205

1

Introduction: the leaf and the pressures it faces

Having evolved for the difficult task of capturing carbon dioxide—a gas that is far less abundant than argon in the air—leaves are the port of entry of carbon into the terrestrial biosphere. But carbon fixed in leaves is, along with other valuable nutrients, a potentially vulnerable food source for other organisms. If an animal does not seize the opportunity to feed on foliage, it will have to attack other less accessible parts of the plant such as bark-covered stems or soil-embedded roots, or it will have to await relatively brief periods of seed production. Alternatively, the animal must kill or parasitize another animal or it will have to compete with micro-organisms to feed on dead tissues. Virtually all terrestrial life relies directly or indirectly on the transfer of carbon from plants in the first trophic level to organisms in higher trophic levels. Yet the immobility and seeming pacificity of the leaves, together with their great abundance on earth, begs the question of why most plants that survive beyond the seedling stage die and decay after reproduction rather than being consumed alive by herbivores. Part of the answer is that obligate leaf eaters have to be highly specialized animals. We are not among them.

It is probable that our ancestors ate much more plant material than most of us do today, and that this included raw foliage. But we typically eat rather small quantities of leaves from a very limited range of plants and, for most of us, however vegetarian we are, the very thought of trying to live on a leaf-only diet for even one week is dismal. Underscoring this is the fact that none of the world's ~12,000 species of grass is used in the leaf stage as a food staple for humans; with remarkably few exceptions the same can be said of the leaves of trees. A similar difficulty seems to confront many vertebrates in the world's great tropical forests where even observing animals can be difficult—there is simply too much greenery. But if leaves, the most abundant organs on earth, provided a more accessible diet, life would be different. Perhaps giant flocks of birds would descend on to the tops of the trees in forests throughout the world to strip off their foliage. What stops this happening? Why is so much of the world green?

Rather than examining how leaves confront environmental extremes, this book concentrates instead on leaf survival in the face of biotic pressures. The focus is on herbivory and particularly on folivory—feeding on living leaves. Before looking at the organisms that are specialized to do this it is necessary to look at the global scale of this phenomenon. How big is the vegetation pool

Published 2014 by Oxford University Press.

on earth and what proportion of plant material escapes herbivory and, instead, is eaten as dead tissue by detritivores? Also, which group—invertebrates or vertebrates—is likely to have exerted the most selection pressure on leaf defences? The answers to these questions can only be approximate, but they can at least be addressed in light of the carbon cycle.

Herbivory and the terrestrial carbon cycle

Recent assessments of the terrestrial limb of the carbon cycle reveal the immensity of its scale. At a minimum, a net 60 Gt (60×10^{15} g) of carbon is fixed by land plants each year, and standing vegetation is a reserve of about ten times this quantity of carbon (Houghton, 2007; Welp et al., 2011). These quantities attest to the importance of terrestrial photosynthesis, but they can also be seen as the success of plants in protecting themselves. The question is what happens to the carbon that is captured by plants? How much is eaten when the plant is still alive, and how much is left to be eaten from dead plants?

In the terrestrial limb of the carbon cycle as it is usually represented, the ~60 Gt of atmospheric carbon per year that is fixed into vegetation is a net gain. Indeed, almost twice as much carbon is captured by plants but about half is then lost through respiration that returns carbon dioxide to the air. Strikingly, in most representations of the carbon cycle (e.g. Houghton, 2007) the 60 Gt of carbon that is fixed into the vegetation pool then flows again out from litter. It is from this dead plant material that the carbon is released back into the air by decomposition (only a small increment is added to the soil each year). In this scheme of things, any indication of the impact of herbivory is notably absent: plants complete their life cycles, then just die. Their carbon is then returned to the atmosphere through a combination of microbial degradation and ingestion by detritivores. This is the typical way in which the terrestrial carbon cycle is represented. Furthermore, the presence of so much vegetation on earth led the ecologist Paul Feeny to famously call herbivory 'a conspicuous non-event' (Feeny, 1975).

Common sense tells us that plants dominate the above-ground biomass of most humid terrestrial environments. On the other hand, much of animal life on land depends directly or indirectly on leaf-eating organisms capturing the carbon that plants have harvested. Is herbivory really a non-event or is its impact just difficult to estimate? The problem is that herbivory is hard to quantify. First, much herbivory is focused on young tissues and thus affects a plant's future growth potential rather than simply removing static plant mass. Second, the amount of damage due to herbivory that will impact the future growth of a plant also depends on its tolerance to being damaged. In addition to this is the fact that plants are distributed very unevenly. Their density per square unit of earth surface differs from almost zero to hot spots in the wet tropics where plant mass is remarkably high, reaching ≥400 t per hectare (Saatchi et al., 2011). Complicating the matter still further is the fact that herbivores are also unevenly distributed in nature. Finally, the extent of below-ground herbivory is largely unknown and is expected to have impacts on the growth of aerial plant parts (Crawley, 1983).

Not surprisingly, all this makes it very hard to extrapolate data to global scales and to estimate what percentage of living plant material is consumed by herbivores each year. Nevertheless, there are some patterns including evidence for increased rates of herbivory nearer to the equator (Coley and Barone, 1996; Schemske et al., 2009). For example, in broad-leaved forests in temperate regions it was estimated that an average of 7.5% of leaf surface was removed by herbivores, and this loss of leaf tissue increased to 10.9% in tropical regions (Coley and Barone, 1996). M.J. Crawley used very extensive literature values from observations such as these to estimate an average of 10% of primary production being lost to herbivory (Crawley, 1983). More recent estimates of plant productivity consumed in terrestrial environments, also based on recorded rates of herbivory, are nearer 18% (Cyr and Pace, 1993). But 'the danger is that once a figure is obtained (usually after great effort), it is likely to be taken too seriously.' (Crawley, 1983, p. 9).

The relative contribution of vertebrates and invertebrates

With Crawley's warning in mind one can make a rough estimate of the extent of herbivory in a different way. Instead of using a meta-analysis of observations of herbivore damage to plants, values from agriculture can be used to look at the problem from a different perspective. The primary reason for this is that humans have extended their farming activities over a vast range of latitudes and across many habitat types so that values represent broad swathes of the earth. Furthermore, the agricultural literature might allow alternative estimations of how much carbon in plants may be eaten while the plant is still alive. For example, this literature contains useful values for how much meat is produced worldwide, of how much agricultural productivity is lost to insects and on soil erosion rates, etc. Keeping in mind that estimates indicate that ~18% of primary productivity is likely to be eaten alive by herbivores (Cyr and Pace, 1993) one can also ask the question of what is the relative impact of herbivory by invertebrates and vertebrates?

Landmass can be divided into uncultivated land, surface that is used for grazing, and land on which crops are cultivated. These represent about 58%, 27%, and 15% of land surface respectively (Asner and Archer, 2010). Of the carbon fixed by crops and on rangelands only ~2 Gt ends up being consumed by livestock (Asner and Archer, 2010; Imhoff et al., 2004). Could this estimate, from exploitation of 42% of the land surface, be extrapolated to include herbivory in nature? After all, throughout the world humans have replaced wild animals with their livestock—and continue to do so. In order to estimate how much living plant carbon might be consumed by vertebrates on a global scale, one can make the simplifying assumption that vertebrates remove primary productivity at similar rates in rangeland and from crops as they do on the 58% of 'uncultivated' land surface. Extrapolating the value of 2 Gt of carbon taken from crops and range to the larger surface of non-agricultural land gives an additional 2.7 Gt loss from vertebrate herbivory. Then, combining these two values we arrive at 4.7 Gt consumed each year by vertebrate herbivores.

The impact of invertebrate herbivores must now be factored in. It has been estimated that there is up to 15% yield loss due to insect pests for cultivated habitats in North America (Losey and Vaughan, 2006). Making the assumption that similar losses due to insects occur worldwide, and taking a roughly median value from previous estimates of herbivory (Crawley, 1983;

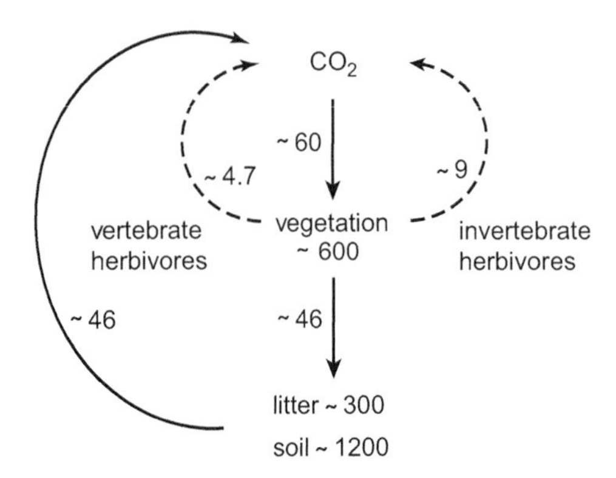

Figure 1.1 Herbivory in the context of the carbon cycle. Rough estimates of the impact of herbivory on carbon flow through the terrestrial carbon cycle are shown. The numbers are billions of tonnes (Gt) of carbon and the overall scheme is from Houghton (2007) and Welp et al. (2011). Added to this are speculative estimates derived from the agricultural literature that attempt to show (dashed lines) how much primary productivity is consumed each year by vertebrate and invertebrate herbivores. The relative impact of vertebrates and invertebrates is expected to differ between ecosystems and between seasons.

Coley and Aide, 1991; Cyr and Pace, 1993; Coley and Barone, 1996; Crawley, 1997) we can use this 15% value for uncultivated land too. This means that 9 Gt (15% of 60 Gt) of fixed carbon might be eaten by invertebrates. If this value is added to the value for vertebrate herbivory then we arrive at ~22% loss of primary production due to combined vertebrate and invertebrate herbivory on a global scale, equating to a little over 10^{10} t of carbon consumed by herbivores per year. This very rough estimate is shown in Figure 1.1.

To summarize, the literature suggests that, at a maximum, each year ~20% of overall plant primary productivity is consumed by herbivores. It would be imprudent to try to extrapolate any further in an attempt to estimate how much of this is specifically the result of leaf eating. As the years go by, estimates for plant consumption by herbivores may creep upwards as we learn more about factors such as below-ground herbivory and ant feeding on extrafloral nectaries, etc. (Crawley, 1997; Coupe and Cahill, 2003). In conclusion, whereas most plant material is decomposed after death rather than being eaten alive, the impact of herbivory is globally significant. Rather than being a 'conspicuous non-event', herbivory is better seen as an inconspicuous event. It is, however, more the exception than the rule to find whole plant communities that have been devastated by herbivores and most of the world's great forests remain green, succumbing more readily to human activity than to herbivores. Something stops the advance of these animals.

The range of leaf-eating organisms to be considered

Having looked briefly at the scale of herbivory it is necessary to define which of the many organisms that attack leaves will be considered here. A line has to be drawn in order to focus on particular defensive traits and on the attack strategies that may have led to their evolution. Figure 1.2 shows where this line is drawn. Only defences against arthropods, molluscs, and

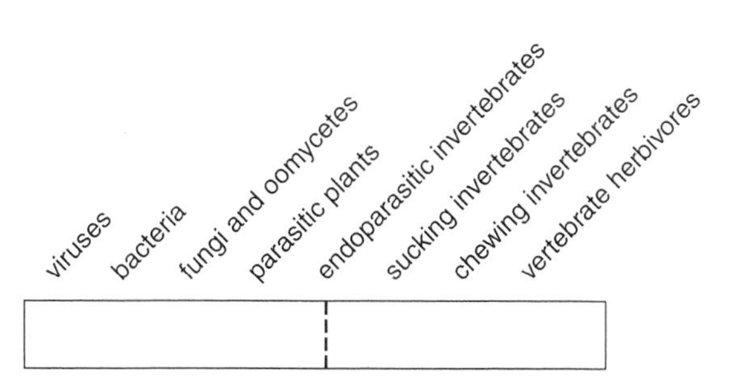

Figure 1.2 The spectrum of leaf attackers considered. Plants are attacked by a broad range of pathogens, parasites, and herbivores; plant defence responses can be divided very roughly into antiviral and antimicrobial and antiplant defences to the left of the dashed line, and antiherbivore defences to the right. Only the organisms to the right of the dashed line running vertically through the box are considered herein.

vertebrates are discussed at any length. The leaf's responses to viruses, countless small organisms from bacteria to fungi, to protozoa, nematodes, and parasitic plants are excluded from discussion. Here the focus is above all on folivores that can be seen readily—those that attack the leaf from outside rather than hiding themselves within its tissues. These are among the most numerous organisms on earth. Each individual plant in nature is, on average, home to more than one of these herbivores.

Herbivores outnumber plants

With few exceptions, each time a plant species is examined carefully it is found to be colonized by a range of invertebrates. For example, a study in New Guinea found an average of 33 beetles, 20 orthopterans, and 26 lepidopteran species per tree species. The bulk of the herbivores were specialists, animals that feed on one or more genera from within one plant family (Novotny et al., 2002). Even more remarkably, in the temperate climate of upstate New York, Root and Cappuccino (1992) found more than 138 different plant-eating insect species associated with goldenrods (*Solidago altissima*, Asteraceae). Of these, seven insect species were considered common on these plants, their biomass reaching up to 0.1% of leaf biomass (the record was held by a beetle). In terms of total insect mass associated with the plants this rarely exceeded 1% of the leaf mass. This pattern of finding many herbivores on individual plants is repeated over and over again. Not surprisingly, the sheer number and diversity of invertebrate folivores pose a defence logistics problem for the plant. As illustrated in Figure 1.3, invertebrates feed on leaves in many different ways and can target nearly all cell types in a leaf. So should the leaf place most of its chemical defences in the epidermis? Could structural or chemical barriers be positioned within a leaf to block organisms or parts of organisms entering its tissues? Before looking at these types of questions we can first pass briefly over the four main branches of the phylum Arthropoda: Hexapoda (Entognatha and insects), Myriapoda (millipedes, centipedes, etc.), Crustacea (woodlice, crabs, etc.), and Chelicerata (spiders, scorpions, mites, etc.). Where in each subphylum did folivory evolve?

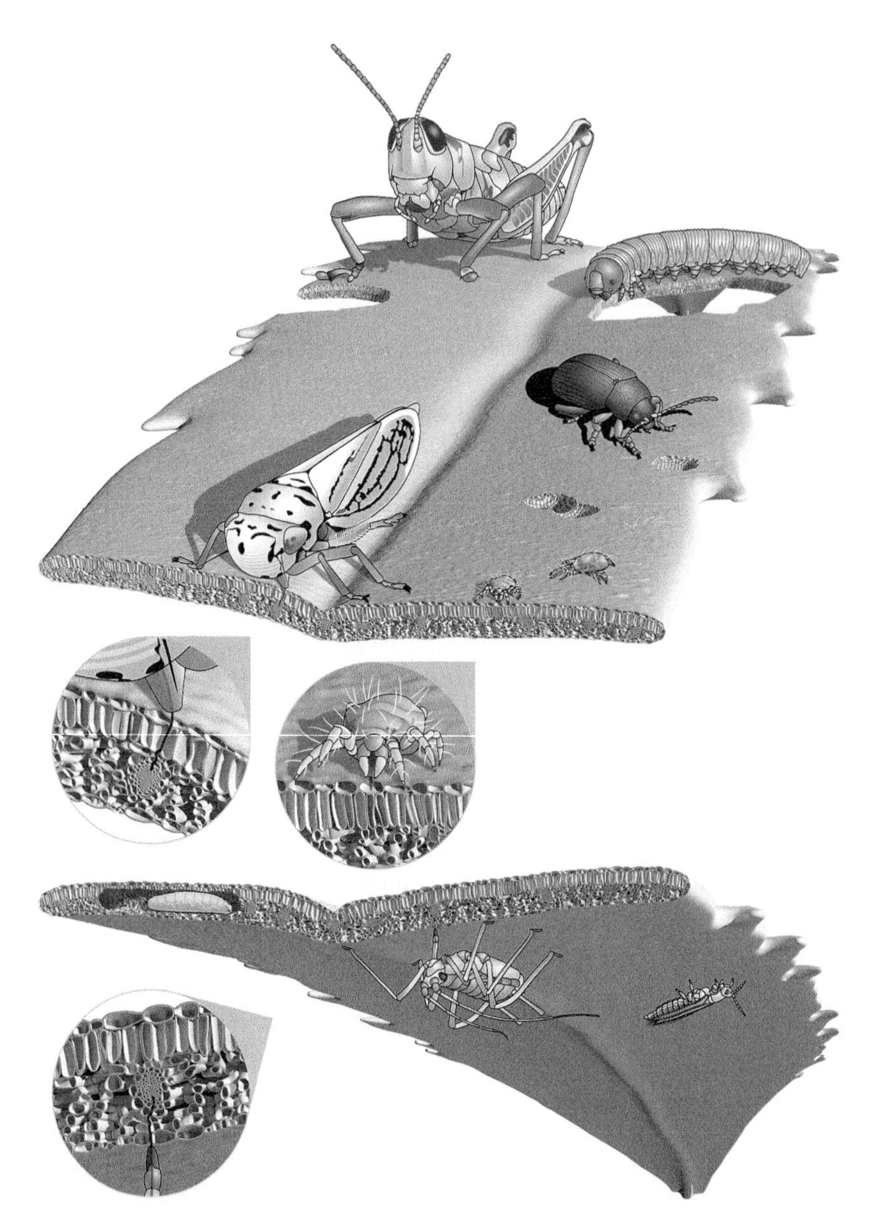

Figure 1.3 Almost every leaf cell type is targeted by selective herbivores. Invertebrates are represented clockwise from the top on the upper leaf surface by an orthopteran (grasshopper), a hymenopteran larva, a coleopteran (flea beetle), two mites (Acari), and a hemipteran (leafhopper). Insets show the mites and the leafhopper feeding on mesophyll and xylem cells respectively. On the lower leaf surface are an aphid (Hemiptera) and a thrips (Thysanoptera) with a lepidopteran leaf miner larva within the mesophyll. The lower inset shows the aphid stylet in phloem tissue. The arthropods shown are not drawn to scale. Not shown are many other types of invertebrate herbivores (e.g. molluscs). Image: T. Degen.

Hexapod herbivores: the dominant roles of insects

Nothing compares to the cloud of noise produced at sunset in wet-season tropical forests by insects as they convert energy from plants to sound. Mostly made by large herbivorous insects, this sound represents only the tip of the iceberg. At the same time, innumerable smaller insects are at work silently, some hidden deep within the tissues of plants. Whereas miners and gallers feed from inside the leaf, the big five herbivore-rich insect orders that (mostly) feed from outside the plant are beetles (Coleoptera), butterflies and moths (Lepidoptera), grasshoppers and locusts (Orthoptera), thrips (Thysanoptera), and aphids, whiteflies, leafhoppers, etc. (Hemiptera) ('CLOTH'). We know more about leaf defence mechanisms against these insects than we do about smaller but no less interesting orders such as Phasmatodea: stick- and leaf-insects. The bulk of the insects in the 'CLO' orders feed by chewing—they injest solid food. This is not the case for the 'TH' orders which feed by piercing and sucking. In discussing leaf responses to attack, most emphasis is placed on insects that chew leaf tissue from outside the leaf. But some of the insects that feed on liquid meals are so highly adapted to feeding on individual leaf cell types that they serve as good examples of the extreme pressures that can be placed on foliage. Very little of this book is devoted aphids and to plant defences against them, but they can be used here to emphasize the sorts of pressures they place on cells in leaves. Subsequently, insects that feed within leaves, leaf miners in particular, are used to illustrate the fact that herbivorous invertebrates frequently if not universally manipulate leaf physiology.

Sucking insects and the stylet

Thrips and hemipterans represent a large proportion of piercing–sucking insects. In aphids the stylet penetrates the host cuticle and, depending on the aphid species and the host plant, then either breaks through cells or bypasses their interiors, often weaving its way through the pectin-rich middle lamellae between cells. The stylet itself is a complex tubular structure that has both a food canal and a separate salivary canal through which the aphid secretes (mostly unknown) substances to facilitate stylet ingress and to dampen the plants' defence response. Once a phloem sieve tube is located, the stylet pierces the cell wall and allows the pressurized sucrose-rich contents to flow into the insect. Feeding aphids get far more sucrose than they need, so they void most of it in order to get enough of the lower abundance nitrogen-containing solutes in the sap. Periodically, certain aphid species are known to withdraw their stylet from the phloem and insert it into the xylem from which they can also take in fluid. This ability of the aphid to correctly pinpoint individual vascular cells is remarkable, since not only are these cells difficult to access but also phloem cells are exquisitely sensitive to intrusion.

There are other insects, e.g. froghoppers (which, as hemipterans, are aphid relatives), that feed most heavily on the xylem. In addition to the robbery of valuable nitrogen this also affects the plants' water relations and these insects can have a comparatively large impact on the health status of the host. This is exactly what was found in insects feeding on goldenrod. Whereas phloem-feeding aphids had no measurable effect on plant growth, common froghoppers (*Philaenus spumarius*) reduced growth significantly (Meyer and Root, 1993). It is not completely clear why an organism that feeds from the xylem should cause such reduced plant growth but the effects of feeding invertebrates on plants are often not simply due to mechanical damage.

The manipulation of leaf cells by miners and gallers

Leaves are not only attacked from the outsides. Numerous moths, flies, sawflies, wasps, and beetles are leaf miners. Female miners either lay their eggs in the expanded or expanding leaf by puncturing through the epidermis, or they deposit their eggs on the leaf surface, and the larvae themselves bore into the leaf. It is often possible to guess what type of insect has made a leaf mine just by looking at the gallery. Mining flies, for example, tend to make long tunnels that become wider as the insect grows. Lepidopterans, on the other hand, often construct 'tentiform' mines that fill entire sectors of leaves between leaf veins. In some of these insects the first instar larvae start out as sap feeders only to become chewing insects at later instars. The galleries that they create are carefully organized and the frass these insects produce is often carefully deposited in latrines. But staying in a single leaf and being in such long duration and close proximity to plant cells is potentially dangerous for the miner. Molecules from their bodies or from their secretions are sometimes detected by pattern recognition receptors in plant cells. Moreover, in some cases cells in the vicinity of these insects undergo a localized hypersensitive response leading to tissue dehydration and isolation of the miner (Fernandes, 1990). But from the plant's point of view an even more extreme defence strategy is to shed the entire leaf, as happens in sand live oaks (*Quercus geminata*, Fagaceae) on the Florida Panhandle. When the leaves of this tree are attacked by a specialist lepidopteran leaf miner (*Stilbosis quadricustatella*) they are often simply shed. So why is this strategy not used by all plants, and why do miners remain as such successful leaf eaters? The answer is that they manipulate their host leaves.

Leaf miners in seasonal climates run the risk of being shed with the autumn leaf fall, before they are mature enough to exit the leaf. So their only hope is to delay leaf abscission in order to maintain their living conditions. In the late autumn in the northern hemisphere, when the leaves of trees are falling, one often finds leaves with prominent 'green islands' (Figure 1.4) of non-senescent lamina within which the individual larvae feed until their development is complete. The compounds that cause these living domains to be maintained in an otherwise dying leaf are not well characterized, but some are produced by the miner's gut bacteria rather than from the miner itself (Kaiser et al., 2010). Green islands exemplify the common and elegant strategies used by invertebrate herbivores to manipulate leaf function but similar, intricate relationships exist between plants and gall-making animals—animals that coax the plant into making structures that it would not normally make. There are more than 100,000 arthropod species that do this, especially among the mites, aphids, wasps, thrips, and flies, but also including some beetles and moths. (Curiously, the majority of these invertebrates only cause galls on dicotyledonous plants; Espirito-Santo and Fernandes, 2007; Redfern, 2011.) It is therefore a problem for the leaf that its physiology may

Figure 1.4 Herbivores can manipulate leaf physiology. Leaf-mining insects risk being isolated in senescing leaf tissue or being shed with leaves in the autumn. As a strategy to complete their life cycles before this happens, many miners either delay leaf fall and cause the formation of green islands in which leaf tissue remains healthy while the rest of the leaf senesces. The leaf shown is from hornbeam (*Carpinus betulus*, Betulaceae) and was probably parasitized by the larva of a pygmy moth (*Stigmella* sp., Lepidoptera). (See Plate 1.)

be manipulated, and the ability of invertebrates to do this is very widespread. Moreover, it is increasingly clear that chewing insects also interfere with leaf physiology, although in many cases what they do is subtle and only becomes apparent through analyses conducted at the molecular level. But whereas this section has emphasized that herbivores can manipulate the physiology of leaves, much of the rest of the book will attempt to show how leaves manipulate herbivores.

Chewing insects

Leaf-chewing insects—like many herbivorous beetles, butterfly and moth larvae, as well as grasshoppers and locusts—all use sharp mouthparts that tend to leave very clean cuts. This has its advantages for the insect. Above all, stealth feeding minimizes the activation of plant defences. But holes in leaves can at least provide useful visual information to many predators and they are also of value to biologists. Looking at them shows us immediately that chewing insects either avoid damaging veins or, in some cases, sever them deliberately. This can be very informative to the naked eye: the sizes, shapes, and positions of holes in leaves often indicate where there are high concentrations of defence chemicals, fibres, or latex-containing cells. Yet more information is easily available from what passes through the insect—although this requires molecular analyses. Chewing insects usually ingest the entire contents of plant cells including extracellular matrices. After digestion, the residue, which is rich in partially digested cell wall material, is evacuated as frass and this is easy to collect and analyse.

The other hexapod herbivores

The huge subphylum Hexapoda in which insects are classified also contains the Entognatha, which comprises three ancient orders: the springtails (Collembola), the two-pronged bristletails (Diplura), and the 'coneheads' (Protura). These groups did not evolve as efficient herbivores although there are at least some leaf eaters among the springtails, an often-cited example being the lucerne flea (*Sminthurus viridis*), a generalist that rasps cells off the lower (abaxial) leaf epidermis of dicotyledonous plants (Shaw and Haughs, 1983). The possibility that there are more herbivorous springtails has been raised by the observation that one species can feed on the living roots or root hairs of grasses (Endlweber et al., 2009). On two occasions I have observed springtails apparently feeding on mosses, plants that are also fed on by some insects (e.g. Haines and Renwick, 2009).

Pressure from non-hexapod arthropods: the mites

Millipedes and centipedes have not evolved the ability to eat living leaves, most being feeders on other animals or on dead and decaying plant tissues; and the Crustacea has very few members that could be considered as folivores. Similarly, with very few exceptions, the spiders and scorpions have also not evolved as herbivores. By contrast with these groups, herbivores do abound within the arthropod order Acari: the mites. This ancient and successful invertebrate order poses a severe threat to leaves, in part because of the great capacity of many mites to detoxify a broad range of plant chemicals. This is a feature of pests such as the well-known two-spotted spider mite (*Tetranychus urticae*), a species that can feed on more than 1000 different plants (Grbić et al., 2011). But life for these invertebrates is by no means trouble-free. Herbivorous mites have enemies in the form of predatory mites and, furthermore, leaf surfaces have often evolved topologies that clearly favour the predators.

Molluscs

The big appetites and the extraordinary capacities of many terrestrial gastropods to survive periods of drought, heat, and cold help to make them among the most redoubtable of all invertebrate folivores. Using hardened radulas slugs and snails scrape at plant tissues. Even the giant coriaceous surfaces of plants like *Gunnera* (Gunneraceae) are no match for these animals, some of which attain sizes approaching those of the smallest vertebrate leaf eaters. While most of us are familiar with small gastropods such as garden snails (*Helix aspersa*), giant slugs such as the spectacular banana slug (*Limax cinereoniger*) from Europe can reach up to 25 cm from mouth to tail, and the bodies of giant African land snails (*Achatina* spp.) are nearly as long. How can any leaf defend itself from such attack? We owe a good part of what we know about this to a classic monograph entitled '*Pflanzen und Schnecken*' (Stahl, 1888), one of the first works to recognize the importance of leaf chemicals in providing defences against herbivores. Today we know that plant chemicals are of primordial importance in defence not only against molluscs but against most herbivores.

Selection pressures and the specificity of defence responses against invertebrates

Insects, mites, and molluscs are both the most numerous and the most diverse groups of plant attackers. Most herbivorous insect orders contain an overwhelming proportion of specialists—often in the range of 60–90% per family (Schoonoven et al., 2005), the rest are generalists, feeding on members of more than one plant family. This is critically important in understanding leaf defence mechanisms and what exerts selection pressure on defences is likely to be suites of insects with similar feeding styles, rather than individual insect species (Maddox and Root, 1990). Even having had the chance to listen first hand to Richard Root, there was a lingering belief in my mind that plants would be able to 'tell the difference' between numerous unrelated insects, even if they had the same feeding mode. To investigate this, my own laboratory made tests of lepidopteran insect-induced gene expression patterns in the leaves of *Arabidopsis thaliana* (Brassicaceae, henceforth referred to as *Arabidopsis*). We compared the plant's gene expression response to feeding of the larvae of the Egyptian cotton leafworm (*Spodoptera littoralis*) with its response to larvae of the small white butterfly (*Pieris rapae*), two chewing insects that are not closely related. The outcome was clear. The overall pattern of gene activity in the leaves was similar (Reymond et al., 2004). A little later, with the benefit of several years' hindsight, the result was not surprising. But if we had compared the feeding of an aphid and a caterpillar (insects with very different feeding styles) on *A. thaliana* the two gene expression profiles in the plant would have been different (De Vos et al., 2005; Bidart-Bouzat and Kliebenstein, 2011). This comes back to the fact that since insects, mites, and molluscs use such a variety of attack strategies the leaf needs a broad range of defences. Furthermore, if invertebrates with their great biological diversity and their clever ways of manipulating leaf physiology were the only folivores on Earth, it would seem an almost impossible challenge for any leaf to survive in nature. But vertebrate herbivores must now be added to the picture.

Vertebrate folivores

Following on from the discussion of invertebrate folivores, the goal here is to concentrate on the vertebrate groups that make leaves the major part of their diet. The emphasis will be on the leaves of land plants, not those growing under water. However—and rather surprisngly—some fully aquatic animals eat the aerial leaves of plants.

Ectotherms: amphibians and reptiles

Cold-blooded leaf-eating vertebrates do not push far into cool climates, although animals living in water might in theory have an advantage over terrestrial leaf eaters because of water's thermal buffering capacity. However, only a few amphibians are known to feed to a limited extent on plants (e.g. Das, 1996) and none is known to feed exclusively on leaves. Concerning reptiles there is curiously little tendency to herbivory in freshwater turtles and terrapins, whereas the majority of land tortoises are highly folivorous throughout their lives. This group includes members that eat a significant amount of grass, such as the desert tortoise (*Gopherus agassizii*) of the Mohave and Sonoran deserts and the gopher tortoise (*Gopherus polyphemus*) of southeastern USA. In the Old World, Europe included, tortoises are the principal extant reptile group that feeds heavily on leaves.

Outside the Old World, however, other reptile groups contain many folivores. Found in Central and South America, the green iguana (*Iguana iguana*) is the best known of the iguanan leaf eaters. Like other herbivorous lizards, these iguanas use hindgut fermentation and they have a caecum teeming with nematodes that presumably contribute to digestion during the slow passage of food through the reptile (van Marken Lichtenbelt, 1993). Other folivorous iguanas include Galapagos land iguanas (*Conolophus* spp.), the desert iguana (*Dipsosaurus dorsalis*) in the Sonoran and Mojave deserts and elsewhere, and chuckwallas (*Sauromalus obesus*) also from the Mohave and Sonoran deserts, for example. In the Agamidae all of the nearly 20 species of burrowing 'spiny-tailed' lizards (*Uromastyx*, found in Africa to parts of Asia) are highly herbivorous, and some whiptail lizards (*Cnemidophorus* spp.) in the Caribbean can consume large amounts of leaves (Zimmerman and Tracy, 1989; Dearing and Schall, 1992). Reptile-specific plant defences surely exist.

Herbivorous fish

Not surprisingly, the list of fish best known for eating leaves is skewed towards those that are popular with aquarists. For example, some cichlids, e.g. the green chromide from southern India, can feed on aquatic macrophytes. A fairly herbivorous characid from the New World is *Brycon petrosus* from Panama and one can include silver dollar fish (*Metynnis* spp.) from South America, and some headstanders (*Abramites* spp.) from the neotropics. But it is probably among the cyprinids that folivory is best known, the grass carp (*Ctenopharyngodon idella*) from East Asia being a good example. This fish begins its free-swimming life as a predator, then transitions progressively to feeding on increasing amounts of plant matter. Fish that feed on aquatic plants are known but it came as a surprise to me that fish could exert pressure on leaves that grow above the waterline.

From a pirogue paddled slowly 3 km downstream of the Kongou falls on the Ivindo river in northern Gabon, we saw some vegetation being disturbed by what appeared to be a rat or an otter. Although we paddled as silently as possible towards the vegetation the creature disappeared, so instead we inspected vegetation above the water-line. In the first area we observed, all of the plants that showed signs of damage were dicots: an *Impatiens* (Balsaminiferae, possibly, *I. irvingii*), a *Ludwigia* (Onagraceae), and two species in the Asteraceae one of which was an *Eclipta* and the other an *Erigeron*-like plant. These plants were growing out from the riverbank in rafts over water at least 3.5 m deep. The damage to the plants was often at the leaf tips but sometimes whole leaves had been removed. We suspected that the animal was a fish and made plans to capture it. The following evening my guides returned with what they called *le capitaine*, a generic name often used in francophone Africa for good-tasting fish. Remarkably, when dissected, the entire digestive system of this cyprinid (probably from the genus *Varicorhinus*) was packed full of leaf material among which swam small white nematodes. Not a single fragment of any invertebrate was seen. This specimen, a fish about 25 cm long, had apparently been feeding itself exclusively on leaves.

Soon after on the same river we observed a more remarkable sight—a shy and darkly coloured fish with prominent eye sockets feeding on the leaves of a large grass (*Echinochloa* sp.). This process required a combination of wiggling vigorously into the grass in order to bend its stems towards the water followed by tugging at the leaves and occasionally jumping to bite them. Tooth marks, each about 3 mm long, were left on the leaves; however, we were unable to catch this fish. A candidate would be a characid, perhaps one related to an African fish called *Distichodus rostratus*, sometimes known as the 'grass eater.' Although generally thought of as feeding on aquatic macrophytes, grass eaters may really have earned their name from eating grass. But the ability to jump to bite leaves is unexpected, and makes these fish the botanists's equivalent of the insect-catching archer fish. This kind of fish and the *capitaine* are likely to be dry-season leaf eaters, feeding on the leaves of terrestrial plants during a season in which there are few fruits, seeds or insects available and when there is likely to be maximum pressure on aquatic macrophytes. They illustrate the lengths to which vertebrates, whether warm- or cold-blooded, will go to exploit leaves.

Endothermic vertebrates

Warm-blooded herbivores at high elevations or high latitudes can put year-round pressure on leaves. Many birds and mammals show remarkable adaptations used to extend their already great capacity to forage under extreme conditions. For example, in winter in the European Alps the alpine ibex reduces its core body temperature and metabolic rate, slows its heartbeat from a summer pace of 90–100 to a winter pace of 40–50 beats per minute, and uses any warmth it can get from the sun to maintain its body temperature (Signer et al., 2011). Together these adaptations allow the ibex to exist on a meagre diet in a hostile winter environment. At the same time, some birds that have broader summer diets use the leaves of evergreen trees to help them survive cold periods. For example, some grouse species have the ability to sustain themselves on the needles of conifers during the winter months. In the presence of warm-blooded animals, evergreen plants may need to be even better defended in winter than in summer when a larger choice of other plants is available and competition for their foliage is reduced. Leaf defences are sometimes seasonal.

Birds

A broad range of birds can eat buds or leaves, some using them as the principal part of their diet. The Anatidae family (ducks, geese, and swans) causes significant impacts to vegetation on several continents and, within this family, the geese are the most folivorous group. Where they graze on grasses and sedges their impact can even outweigh that of mammalian herbivores (Cargill and Jefferies, 1984). Geese, of course, are not the only birds that damage crops. Various corvids, thrushes, and pigeons do this too. Woodpidgeons, for example, can strip leaves off plants, leaving only the main veins (Murton et al., 1966) and many other large pigeons throughout the world will feed on young tree leaves. In Africa, the mousebirds (Coliidae) can strip buds and sometimes even large leaves from plants. Some of the Gruiformes, an order that includes cranes, coots, and the New Zealand takahē, are also partially folivorous, and another New Zealand bird, the strange and flightless owl parrot or kakapo, eats a considerable amount of lycopod and fern material along with the leaves of angiosperms from the Ericaeae and other families.

A feature of many of the best avian leaf eaters is that they are either poor flyers or cannot fly at all. This is the case for three ratites that eat considerable amounts of grass, also consuming other vegetable and animal matter. The ostrich often accompanies antelopes and zebras in mixed herds and a similar phenomenon occurs in eastern South America where greater rheas graze near to guanacos. In Australia the emu takes on the role of the most folivorous of the extant ratites. But where are the flocks of birds that should fly down to the crowns of trees and denude them of their foliage?

Perhaps the closest one comes to an arboreal avian leaf eater is the hoatzin (*Opisthocomus hoazin*), a dedicated folivore found in colonies along some of the warm river deltas in north-eastern South America. Folivores par excellence, hoatzins, like some of their primate counterparts (e.g. the proboscis monkey), are never far from river deltas, habitats that are typically richer in available nutrients than the interiors of forests. These ancient, peculiar, and bulky birds have a diet consisting almost exclusively of leaves, even those known to contain high levels of defensive chemicals, for example the leaves of philodendrons which are rich in phenolic toxins. Hoatzins are the only bird known to use foregut fermentation and in this respect they strongly parallel ungulate ruminants (Grajal et al., 1989; Godoy-Vitorino et al., 2008).

Mammals

As a way of looking at the mammal groups that contain obligate folivores we can ask which groups depend absolutely on leaves for their year-round food supply. Not surprisingly, answering this question is difficult. First, geographically distinct populations of the same species may exploit plants in different ways. Second, where herbivores in cooler climates do not hibernate and do not make food caches, their diet is often seasonal, as is the diet of many animals in warm climates with distinct wet and dry seasons. Nevertheless, making rough estimates for this can be attempted for Africa, a continent that has a diverse and well-documented fauna (Kingdon, 2001). Table 1.1 gives rough estimates for the numbers for

Table 1.1 Rough estimates of the numbers of species of obligate mammalian folivores in Africa

Animal group	**Number of species**	
	Strict leaf diet	Total
Carnivores, insectivores, elephant shrews	0	252
Primates	1	83
Lagomorphs	9	9
Rodents	?0	383
Bats	0	201
Hyraxes	?1	11
Elephant	0	1
Aardvark	0	1
Odd-toed ungulates (equids and rhinos)	6	6
Even-toed ungulates (e.g. antelopes and bovids)	53	90
Total	70 (6.8%)	1,037

the most dedicated folivores among African mammals. Elephants are not included since these, depending on the season, eat fruit and bark, etc., as well as large quantities of leaves. This table represents only the very top of the pyramid, the ~7% of African mammal species that depend on leaves for over nine-tenths of their diet—folivores in the strictest sense. Three-quarters of these are ungulates and most of the rest are lagomorphs. These two groups are also the main extant mammalian folivores in Eurasia and the Americas; macropods only replace them in importance in Australia and New Guinea.

Ungulates and lagomorphs: the dominant mammalian folivores

The ungulates are the unsurpassed vertebrate leaf eaters and over the last tens of millions of years no other group of mammalian herbivores has had such an impact on the vegetation cover of the Earth. The majority of ungulates are leaf eaters and much of their success is related to the development of rumination, a process that allows repetitive processing of relatively large fibrous pieces of foliage. Within the rumen, pieces of intact leaf float on a soup of finer plant material and are then skimmed off and refed to the mouth for further chewing before being re-swallowed. Deer, moose, giraffes and all bovids (antelopes, camels, cows, goats, sheep, etc.) do this, although it has some disadvantages: these animals cannot enjoy the pleasures of lying on their backs to ruminate. Pigs, hippos, rhinos, and equids, by contrast, do not ruminate. Being able to ruminate does not correlate strongly with the ability to feed on particular groups of plants (Van Soest, 1996).

As illustrated for Africa in Table 1.1, and as is the case elsewhere, lagomorphs (pikas, hares, and rabbits) are important folivores in many habitats over huge temperature extremes, from high in mountains to hot deserts. Being small means that lagomorph digestive systems have to be extremely performant in order to access fibrous foods such as grasses; thus the group has developed caecotrophy, whereby the animal produces two types of faeces. One is the commonly seen rabbit droppings which are pellets packed with coarse fibrous matter. The other is reinjested and is rich in microbial protein and essential nutrients. Although there are many leaf defence mechanisms dedicated to destroying plant-derived nutrients in situ in the animal gut, some of these nutrients can be reconstituted in animals with hindgut symbionts. Symbionts add to the odds that seem to be stacked against leaves.

Marsupials

Two-thirds of the world's marsupial species live in the Australo-Papuan region where kangaroos and wallabies (macropods) form the largest group of vertebrate folivores. Most of the macropods (approximately 50 species are strict grazers and they are the equivalents of herds of grass-eating ungulates in most of the rest of the world. Those macropods that include substantially less grass in their diets include pademelons, hare-wallabies, and the swamp wallaby, all relatively small animals. The swamp wallaby is one of relatively few vertebrates that sometimes feeds on the poisonous leaves of bracken ferns (*Pteridium sp.*). Finally, the remarkable tree kangaroos of New Guinea and their counterparts in Australia eat considerable amounts of leaves, as do many possums some of which are obligate folivores. Some of these will be mentioned later along with an unrelated but well-known animal that has a strict leaf diet—the koala. But returning to the ground, there are yet other marsupials that eat leaves—for example, wombats). Clearly, plants in Australia have faced mammalian folivores that differ from those elsewhere in the world, and this allows the continent to be used as a natural laboratory to study the question of whether defence mechanisms used by plants against marsupials might differ from those used against placental mammals. When it comes to defence strategies, are plants in the Australo-Papuan region different from those in the rest of the world? The short answer is, with some exceptions, not much.

Leaf eating in other mammalian groups: primates, rodents, and bats

Although the purpose of Table 1.1 is to highlight the prominent place of lagomorphs and ungulates as leaf eaters, it should not give the impression that only a small proportion of African mammal species put great pressure on leaf defences and that all of this pressure is from ground-based animals. That would be wrong. For example, whereas only one primate is alluded to in the table, the gelada (*Theropithecus gelada*), a baboon relative that eats grass as the single major component of its diet, many African primates from great apes to monkeys eat large amounts of leaves. Among the most folivorous tree-living monkeys in Africa are colobids. Further afield, another of the very dedicated tree-living leaf eaters is the proboscis monkey from Borneo. In the Old World there are also the prosimians. Most of these feed heavily on insects or gums but Madagascar has some very folivorous lemurs. Then there are several langurs in Asia that feed heavily on leaves. In the New World there are the howler monkeys, quite large animals known for their eerie vocalizations as well as for their excellent colour vision. In summary, primates are the most widespread group of arboreal, warm-blooded leaf eaters. They have almost certainly selected for some of the defence-related characteristics of tree leaves in some of the world's warmer forests. Below them, at ground-level in South America, certain large rodents must have also put vegetation under considerable pressure.

South America has the most spectacular of the World's leaf-eating rodents. With their huge range throughout the wetlands of central and eastern South America, capybara eat large quantities of grasses and aquatic macrophytes, and coypus eat a broad range of aquatic vegetation. Other highly folivorous rodents in South America are cavies, relatives of the capybara. Both these types of animal use caecotrophy to maximize the efficiency of digestion, as do lemmings. Leaf eating among rodents tends to decrease with smaller size. Looking again to Africa, most of the ~400 species of murids are herbivores, but there are probably none that feed exclusively on leaves. So when mice or voles in various parts of the world are found eating leaves this merits publication (e.g. Timm and Vriesendorp, 2003). Instead of eating foliage most rodents prefer to eat more energy-rich plant parts.

Finally, given what we know about mice and rats, it is not surprising that no bats are obligate folivores, although some insect-eating bats and some fruit bats have nevertheless developed an unusual way of feeding from leaves. These bats take bites out of foliage and then, using their tongues, squeeze the leaf material against the roof of their mouths. The fibrous debris that remain are then spat out and the liquid contents, a dietary calcium supplement (Nelson et al., 2005), are swallowed. Squeezing the juice out of leaves might seem to be an effective form of folivory, but the fact it has not been widely adopted suggests that this feeding mode has its limitations. Leaves or parts of leaves have to be swallowed and dealt with in toto in the digestive system in order to sustain vertebrate folivores.

In summary, a broad range of vertebrates feed on leaves and they differ greatly in physiology, size, and feeding strategy. Vertebrate herbivores from reptiles to large ungulates not only feed more or less selectively on leaves but they also cause collateral damage to plants. A single moose (*Alces alces*) in Scandinavia can eat 8000 kg of vegetation per year and the damage it causes simply by smashing through the undergrowth is considerable (Persson et al., 2000). Therefore, in addition to specific defence strategies, leaves need general mechanisms that detect and respond to any form of damage whether it is caused by a falling branch or by a large animal. Beyond this, specific defensive adaptations to vertebrate herbivores can be expected. But prior to looking at this it is first necessary to examine the basic structure and the nutritional value of the leaf.

The nature of the leaf

Here, the term 'leaf' is used in a loose sense for the stem-derived photosynthetic blades of 'euphyllophytes', plants represented on today's earth by ferns and by seed plants, the majority of which are gymnosperms and flowering plants. In some sections this definition is extended to include phyllodes, leaf-like organs that are derived from petioles (these are abundant in some floras, e.g. in Australian acacias) and cladodes, flattened branches typical of prickly pear cacti (*Opuntia* spp.). The leaves of angiosperms, gymnosperms and, to a lesser extent, ferns, are the main focus. Cotyledons, organs that have functions somewhat different from those of 'true' leaves, are discussed only briefly, as are microphylls, the leaf-like photosynthetic organs of lycophytes. Leaves of various types started to appear ~370 million years ago, prior to the beginning of the Carboniferous period. This was at a time when atmospheric carbon dioxide levels were falling steeply, so that increasingly larger stomate-bearing surfaces were needed to capture this diminishing gas (Beerling, 2005; Beerling and Fleming, 2007). With the expansion of the leaf blade came the increasing need to defend it.

But how can this be done? For one thing, a notable feature of leaves is that they have evolved a seemingly overwhelming variety of sizes, thicknesses, and shapes. Excluding vascular bundles which are complex arrangements of multiple cell types, the leaf of an oak may be made of only six to eight cell layers and the lamina may have a maximum thickness of <500 μm. The surface:volume ratio of a large expanded oak leaf will be >5000, but the largest leaves, e.g. those of giant palms, can have surface:volume ratios nearly two orders of magnitude greater than this. The larger the surface:volume ratio, the more the risk of damage. In terms of leaf thickness, at one extreme are the filmy ferns (*Hymenophyllum* spp., Hymenophyllaceae) where most of the tissue between the vascular strands in leaves comprises a single translucent cell layer that has combined features of the epidermis and photosynthetic cells. At the other extreme are the thick leaves of succulents.

To complicate matters, leaf diversity on an individual plant is usual if not universal. That is, at any one time, a plant may have several leaf types. First, these can be produced in a developmental series. After the seedling phase, angiosperms pass into the juvenile phase, then through to an adult phase and, finally, to the flowering phase. Leaves produced at each of these developmental phases are often visually distinguishable and they may be retained throughout much of the life of the plant—so that an individual plant may simultaneously bear at least three basic leaf types. Not surprisingly, then, plants may have different defence needs and different defence capacities at each growth phase (Karban and Thaler, 1999). Also adding to the diversity of leaf types on an individual plant is the fact that many trees produce both 'sun' leaves and 'shade' leaves that can differ in cellular structure. Furthermore, some plants produce a series of different leaf types depending on the season. For example, perennial shrubs that grow in climates with dry periods often shed one leaf set to replace them with more drought-tolerant leaves. Leaf diversity is one of the factors that can affect the distributions of small herbivores on large plants.

Other features can be added to the seemingly endless diversity of leaf types. One of these is leaf lifespan which can vary by as much as 320-fold between species (Wright et al., 2004). This and yet other variables make it difficult to generalize about the properties of these organs, but there are patterns—or more specifically, covariance of important parameters—that describe the leaf blade. This has been investigated in the framework of the 'leaf economics spectrum' (Wright et al., 2004; Osnas et al., 2013). For example, the longest-lived leaves tend to be the thickest. Moreover, thick leaves tend to contain less nitrogen and phosphorus per unit of mass than do thinner leaves and this is highly relevant in the context of defence. On the one hand, the longer-lived the leaf, the higher the chance is that it will encounter an attack. On the other hand, the low levels of valuable nutrients in thick leaves reduce the attractiveness of these leaves to herbivores. Finding apparently universal relationships such as these simplifies our understanding of what makes certain leaves attractive to leaf eaters. Can we make similar sorts of generalizations at the cellular level? If so, what types of cells in a leaf are of most value to a leaf eater?

Leaf cell types

The complexity of the leaf can be simplified by regarding the early developmental origins of its cells. From this perspective the leaf of dicotyledons can be divided into three layers each of which is derived from different cell populations in the embryo. These are layer 1 (L1), epidermal cells and their derivatives; layer 2 (L2), the ground tissues situated between the epidermis and the vasculature and including the major photosynthetic cells; and layer 3 (L3), all vascular cells (Figure 1.5). Starting with on the outside, the majority of leaves have distinct upper (adaxial) and lower (abaxial) epidermises. The basic and most numerous cell type in the epidermal layer is the 'pavement' cell, the predominant cell type from which more specialized cells differentiate.

With few exceptions such as *Hymenophyllum*, the leaf epidermis contains stomata, structures typically based on pairs of guard cells. These function to allow the entry of carbon dioxide which is then supplied via the substomatal cavity to the photosynthetic cells. At the same time, they allow the controlled release of water through transpiration; it is estimated that for each molecule of CO_2 that is captured by the leaf, several hundred water molecules are lost. On most cotyledons, stomata are relatively static pores that remain open throughout the day and night. But those on leaves are dynamic. Paired guard cells control the aperture of the stomatal pore, closing it in darkness or when water availability is reduced. Stomata are a first obvious site of entry for the stylet of a small probing herbivore such as an aphid. But for reasons that will be examined, one does not typically find these insects using these openings as points of entry into the leaf.

The epidermal layer of most leaves typically has populations of specialized cells that are less abundant than those of stomata. These are hydathodes and, unlike leaf stomata, they

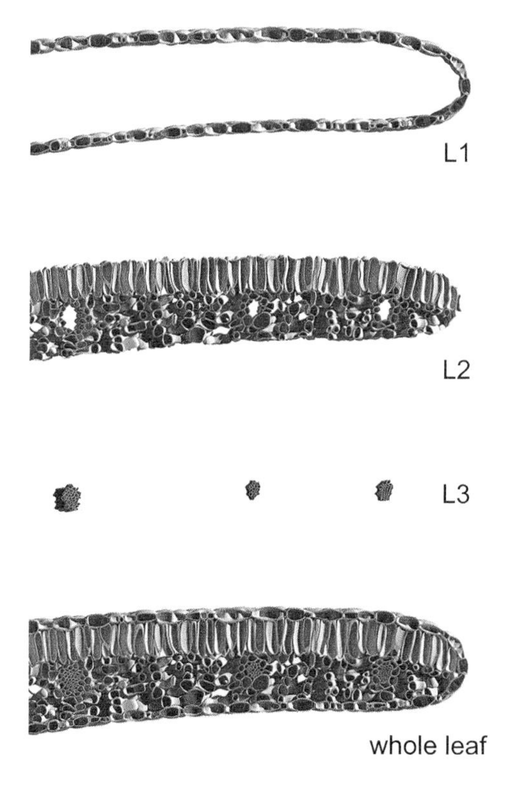

Figure 1.5 Cells in a dicotyledon leaf are derived from three embryonic layers: L1, L2, and L3. These are shown in transversal sections of a virtual *Helleborus* leaf that is reconstructed in the lower image. The leaves of monocotyledons and gymnosperms have similar structures but are derived from two and (probably) one embryonic cell layers respectively. Image: T. Degen.

remain open and release water (and minerals) through the process of guttation. Hydathodes are typically found at leaf tips and leaf edges, especially at the tips of leaf serrations. Like stomata, they are potential points of ingress and they need to be protected. In some plants, part of this protection is provided by a fourth cell type: the trichome. These are primarily specialized defence structures formed of one or more epidermal cells and they have a great variety of structures ranging from single cells to large clusters of cells borne on multicellular stalks. Given their importance in defence these cells will be treated in more detail later, as will other defence-related cells.

Regarding layer 2, most of the upper epidermal layer is usually transparent so that when one looks at a leaf one is often focusing through the epidermis and on to this second layer of green photosynthetic cells. There is enormous diversity in the structure of L2 among leaves. In the majority of species the cells directly below the epidermis are photosynthetic cells and, in terms of volume, these are typically the largest cells in a leaf. More often than not the upper layer of these cells forms a palisade, a cell layer characterized by usually cylindrical cells that are elongated in the plane of leaf thickness. Below these cells are spongy mesophyll cells that also carry out photosynthesis. These latter cells tend to be more rounded than the palisade cells. Being large, photosynthetic cells have the capacity to store significant quantities of defence chemicals.

Finally, organs need vascular systems and the leaf is no exception. Throughout the plant the vascular system is derived from L3 and this layer probably contains the greatest diversity of cell types in the leaf, the best known of which are the xylem and phloem, two largely interdependent cell complexes. In the leaf these complexes and all their associated cells which include the vascular cambium are essentially wrapped in a layer of L2-derived cells—the bundle sheath. This more or less distinct cell layer envelopes the leaf vasculature. Beginning with the leaf xylem, this is typically a collection of cells that range from living, undifferentiated xylem parenchyma through to dead and fully differentiated xylem vessels. Among the latter are protoxylem cells which differentiate while the leaf is still expanding, their spiral spring-like cell wall thickenings allowing these cells to stretch during the completion of leaf growth. Differentiating later as the leaf completes expansion, the metaxylem cells with their complex wall thickenings are thought to be less extensible and they add to the rigidity of the mature leaf. The mature xylem transports water, minerals, and some nitrogen-containing compounds such as amino acids but these cells are almost invariably associated with living xylem cell precursors, some of which display heavily cellulose-thickened cell walls. The function of these cells in leaves is still mysterious.

Phloem is a complex of two main cell types: sieve tubes and companion cells. Phloem sieve cells are among the strangest in nature. They are living but enucleate; they depend heavily on their companion cells but they contain membrane-bound compartments. Their content is a kind of dilute but sugar-rich cytosol—a food source for by many invertebrate herbivores—most notably aphids. Each sieve tube functions together with its companion cell, metabolically active cells that act both to maintain and support the sieve tubes and, through nutrient ztransport, to connect them to the rest of the plant. The sieve cells are connected to one another in the longitudinal axis with sieve plates, perforated structures that quickly become blocked in a calcium-dependent process in response to intrusion. Aphids periodically salivate into the sieve tube to counter phloem occlusion (Will et al., 2007). In summary, small herbivores like aphids and miners constantly monitor the cells they dine on. As will be seen, plant cells constantly monitor herbivore activity.

The leaf as a food source

Having looked briefly at a number of leaf cells we can now ask which cells are the most sought-after by herbivores? It is clear that the majority of cell types in leaves are vulnerable to predation, so generalizing about how much pressure herbivores exert on particular cells would appear to be difficult. All tissue layers in a leaf are preyed on, often very selectively. However, the leaf tissue dominated by photosynthetic cells is where the bulk of the nutrients in the first trophic level are found and it is not an exaggeration to say that L2 (and its equivalent in monocotyledons) feeds the world. Vascular tissue, and in particular phloem, a tissue that receives sugar exported from photosynthetic cells, follows closely behind. But a problem confronting folivores is that much of the carbon in leaves is tied up in hard-to-digest materials such as cellulose and lignin. Carbon fixed in lignin is nutritionally unavailable, and cellulose, found throughout the leaf, is only partially accessible to specialized herbivores and inaccessible to some. But a proportion of the carbon that is locked up in hemicellulose, pectin, and other polymers is more readily digested with the help of symbionts. Over the course of evolution, herbivores have formed partnerships with various bacteria, protozoa, and nematodes many of which help them to degrade these and other carbohydrate polymers. So the herbivores that harbour these organisms, both vertebrates and invertebrates, pose an added threat to leaves—they can exploit a high proportion of carbohydrate and they can get this from most cell types. Aside from the readily exploited and concentrated source of sugar in the phloem there are other pools of sugars and easily hydrolysable poysaccharides such as starch. Starch levels vary considerably during the day, reaching a maximum just before nightfall at which point they can attain levels of tens of micrograms per gram leaf fresh weight. But these are low amounts: what is it about photosynthetic cells that makes them so important in the nutrition of so many leaf eaters, and to so many of the animals that eat leaf eaters?

These cells are where most of the protein in leaves is found. The dominant protein in the leaves of most plants, the carbon fixation enzyme ribulose bisphosphate carboxylase/oxygenase (Rubisco), is a principal source of valuable amino acids. Together, the two protein subunit types that make the functional Rubisco protein may contain up to nearly 40% of the amino acids in a leaf and this is a potentially greater source of energy than a few micrograms of starch. Additionally, a significant proportion (often ≥1% of leaf dry weight) is chlorophyll, nitrogen- and magnesium-rich molecules that are cradled into complexes of chlorophyll-binding proteins within photosynthetic thylakoid. Despite the fact that the ratio of protein:digestible carbohydrate is relatively high in foliage, leaves are usually poor energy sources. If one goes by the calorific content of foods, then to get the equivalent of a 200 g steak, one would have to eat nearly 3.1 kg of nettles (as a particularly nutritious leaf vegetable) or >8.6 kg of lettuce. Nettles and lettuce leaves are eaten by humans. But if grass leaves were analysed for calorific content many would fall below the level of lettuce. Moreover, the energy content of foods as routinely estimated by combustion (and expressed as Joules or kilocalories) is not equivalent to their digestibility. Instead, if one were to assess the true value of steak and leaves as energy sources for humans it is likely that the latter would be an even poorer food source—especially if they were eaten raw. This is because leaves typically contain defence molecules that have evolved specifically to impair the function of the digestive system of leaf eaters. To conclude, whereas the overall content of protein, digestible carbohydrate and lipid is low in leaves relative to many other food sources (e.g. seeds, meat, etc.), the protein content of leaves is where the majority of folivores obtain the majority of their calories. The leaf must put most of its defences into protecting protein and amino acids.

There are yet other aspects of nutrition that go beyond the ability to digest certain components of plants. These relate to the deepest evolutionary history of plant and animal cells and to the fact that it is often easiest to consume an organism closely related to oneself. For example, plants and animals differ in their need for certain ions. Sodium, while dispensable for most plant cells, is a major and indispensable ion in animals. But most plants are poor sodium sources, so herbivores have to be highly efficient at recovering this ion, or they must take time from feeding to seek out sodium sources in the environment. Plants can exploit this, making a herbivore's life more difficult when it comes to retaining sodium (and water). To do this they can produce diuretics and these may be important defence compounds in nature (Dearing et al., 2001). In most habitats obtaining enough salt is an issue for the leaf eater, and evolutionary contingency has placed the leaf in a good situation in this respect.

Many herbivores target young leaves

Whether they are invertebrate or vertebrate, many herbivores are attracted to young leaves and the reason for this is likely to be multifactorial. First, expanding leaves may contain a relatively higher concentration of nutrients than mature leaves because the expanding leaf is a sink that receives photosynthate and nutrients from elsewhere in the plant. Additionally, the vacuoles of cells in small expanding leaves are usually far smaller than those in fully expanded leaves. This means that there is proportionally more cytoplasm with a high protein content to feed on and also that there is a reduced volume of vacuoles in which to store chemical defence molecules. Nevertheless, for all but the smallest herbivores the zones of cell proliferation (meristems) probably add little to the attraction of young leaves; even where extensive meristem tissues are found, such as in the heads of cauliflowers, the tissues are well defended. So what are the other features of young leaves that make them attractive to folivores?

One of the big differences between young, expanding leaves and mature, full-sized leaves is that the xylem is highly differentiated in the latter, that is older leaves contain more lignin than occurs in younger leaves. The lignin content contributes to a leaf's toughness and the toughest leaves tend to be avoided by herbivores (Aide, 1993): they are time-consuming to chew and fibres can slow digestion (Owen-Smith and Novellie, 1982). Several studies on Barro Colorado island in Panama underscore this. Howler monkeys on the island chose the leaves they ate mostly based on protein level and fibre content (Milton, 1979). On the same island, herbivory rates due largely to invertebrates were correlated with leaf toughness and fibre content (Coley, 1983). The young leaves that are eaten so readily are rich in expanding and expanded cells rather than dividing cells or cells that have fully differentiated. So, to a large extent, the physical properties of the leaf help to determine its attractiveness to both invertebrates and vertebrates.

What is leaf defence?

This book emphasizes the extent to which the visual features, cellular structures, and mechanical and chemical properties of leaves have been shaped through pressures from herbivory. This is, of course, not intended to diminish the great importance of the physical environment as the source of selection pressures that have fashioned leaf form and function. Nevertheless, an underlying current in the book is that biological forces have shaped not only leaf chemistry but also much of the cellular construction of leaves. The reach of these biological selection pressures in shaping leaf defence biology has often been underestimated. For example, in the words of the evolutionary ecologist D.H. Janzen, '. . . it has been fashionable to try to understand the spininess of arid-land plants largely in the context of their interactions with the physical environment' (Janzen, 1986). But animals have had an even greater impact on spininess than has climate. Furthermore, while nearly everyone seems to be willing to agree that some insects have evolved to escape predation by resembling dry twigs, there is often a deeper reticence to accept that some perfectly healthy plants also resemble dry twigs to escape predation. In words that are not my own, flowers are beautifully adapted to pollinators but it seems harder to accept that leaves are highly adapted to herbivores. But a failure to examine the hypothesis that plants use ploys such as mimicry to minimize detection by herbivores is problematic. It can lead to treating plants and the animals that feed on them in an independent manner, rather than looking at them as co-evolved partners sharing the same environment and confronting parallel difficulties of living and reproducing in the face of often similar biotic and abiotic pressures. Moreover, many articles on plant defences begin with sentences like 'Plants, being sessile organisms, are unable to avoid attack and need powerful defences.' This reasoning, based on lack of mobility, is not helpful in understanding many of the defence strategies used by leaves. A bird, for example, is just as likely as a plant to be infected by microbial pathogens. To some, the sentence might imply that plants have good basal defences but that they lack rapid responses. But leaves are exquisitely sensitive structures that constantly monitor the environment and their own health status, and active, highly co-ordinated responses to attack are initiated in seconds. Nearly all of these reactions are invisible to the human eye and it is only through working with sophisticated equipment that we can investigate this.

Here, defence is defined as the physical and chemical features of plants that reduce predation and conserve reproductive fitness. This definition is broad. It covers stings, thorns, and trichomes, and extends to the colours and forms of leaves that might help them hide in the environment or might serve as warning signs, and it also encompasses chemicals that plants make in order to deter or repel herbivores. Chemicals and physical defences that act directly on the attacker are known as direct defences—a plant's autonomous contribution to its own defence. But common sense tells us that the numbers of herbivores in any particular environment also depend on the abundance and proximity of their predators. Consistent with this, many plants have evolved mechanisms to reward carnivores, either by providing them with food or accomodation, or by giving them information as to the whereabouts of the herbivores. Together, this is known as indirect defence, that is, the predator provides protection to the plant. Many of the really interesting features of direct and indirect defence mechanisms cannot be seen or sensed easily; they are hidden to the human eye and they are often surprising. But before looking at some of them it is necessary to turn to several plant responses associated with attack but that are not always viewed as defence responses. These are recovery growth (more often called compensatory growth) and tolerance. The two phenomena are often interrelated and they can be important in plant survival in the face of herbivory.

Browser feeding and recovery growth

The overall architecture of many plants can be strongly influenced by herbivory, and what happens in the days and weeks following damage will often depend on whether or not the plant's lateral meristems are activated after removal of apical meristems. Whereas overall architecture can differ greatly between species (e.g. between a willow tree, a buttercup and a vine) the apical meristems in these plants are found in relatively exposed positions and browsers in search of young foliage focus their attention on these regions. If a herbivore bites off apical meristems it may impact the position and growth of new branches. An excellent example of this is the regrowth of tree stumps after beaver damage in North America and Europe. If a willow tree (*Salix* sp.) is felled by a beaver it will usually resprout many lateral branches from its stump. These will later develop into a many-stemmed tree that might not be as attractive to the beaver as the single-stemmed tree it originally cut. But when adequate recovery growth mechanisms have not evolved, plants can be particularly vulnerable to introduced herbivores. An example of this concerns beavers that have been introduced from North America into Patagonia. These animals often fell the southern beeches (*Nothofagus* spp.; Nothofagaceae) which then die because they, unlike their northern counterparts, cannot develop new lateral shoots from the remaining stumps (Anderson et al., 2006). Recovery growth, where it has evolved, is not simply restricted to compensating for lost photosynthetic tissue: it can alter the architecture of plants in ways that are likely to favour their survival in the face of continued herbivory. Related to this, trees that support coppicing probably evolved this as a response to pressure from large herbivores.

Tolerance, grasses, and grazers

Often associated with recovery growth is the phenomenon of tolerance. Again this is generally not considered as a defence, but it is critical to plant survival. Tolerance is defined as the ability to reproduce successfully despite having been damaged (van de Meijden et al., 1988; Strauss and Agrawal, 1999; Juenger and Lennartsson 2000; Kotanen and Rosenthal 2000). This is a whole organism-level definition but, clearly, the ability to tolerate damage is essential to the function of individual leaves. No organ continues to function as well when damaged as does a leaf (Figure 1.6). Nevertheless, in many cases tolerance is built into the entire architecture of the plant. This is the case in grasses.

Fresh grass is the driver of most of the great but dwindling terrestrial migrations, from those of chiru, khulan, and saiga of central Eurasia, elk and pronghorn in North America, and wildebeest in East Africa. The pressure exerted on grassland by animals such as these can be so great that one has to ask how grasses survive this yearly onslaught. Do they use different defence strategies to those of other plants? In terms of chemical or physical defence the general answer is likely to be no, although with exceptions grasses do tend to be somewhat poorer producers of leaf defence chemicals than many other plants. Alternatively, are there fundamental differences between the animals that are adapted to feed on grass and those that eat other types of leaves—browsers? The fibrous nature of grass compared to most other plants requires a relatively large and performant digestive system but not necessarily one that is fundamentally different to that of browsers (Prins and Gordon, 2008). While there is indeed an overall tendency in some vertebrate groups (e.g. antelopes) for grazers to be heavier than browsers, the fact that grazers are often not much bigger than their browsing counterparts suggests that basically similar problems must be overcome in order for the herbivore to be able to digest grasses or entirely unrelated plants. How, then, do grasses survive?

Figure 1.6 Tolerance to damage, a quintessential feature of leaves. This leaf of *Tussilago farfara* (coltsfoot, Asteraceae) from the European Alps has been severely damaged by chrysomelid beetles but is still partly functional. Leaves show more tolerance to damage than most other organs in nature. The leaf was ~20 cm in width. (See Plate 2.)

Clues as to how grass differs from most other plants come from finer details seen in closely related grazers and browsers. Good examples are African rhinos. The overall shapes of the bodies of black (browse) rhinos and white (grass) rhinos are similar, but they differ markedly in the part of the body that comes into contact with the plants they eat. The former is a browser; it has pouting lips. The white rhino, by contrast, has a squared face well adapted to feeding very close to the ground. This apparently small but significant difference in the bodies of these two animals is specifically related to the white rhino's principal food: grasses. What makes many grasses and sedges so tolerant of heavy attack by specialized grazers is, above all, the fact that the crown (apical) meristems that give rise to the above-ground parts of the plant are physically protected by being deeply embedded in the centre of the grass body, often in a ground-hugging position. Grazers, instead of destroying crown meristems, typically prune away older peripheral leaves and this can actually favour new growth and be beneficial for the plant. For example, in tussock-forming species the removal of older leaves by grazing lets in light to the growing centre, promoting vigorous growth. Not only are there direct effects of grazing on the grass. Nitrogen that can otherwise be locked up for long periods in dead leaves is returned in a soluble form to the ground as droppings and urine and, from this, plants can recapture it rapidly. The effect of grazing is easily seen. When grazers are eliminated by fencing, the colour of entire fields can change, becoming whiter due to an accumulation of dead grass leaves. So perennial grasses have body forms that are optimized to withstand damage, to recover from it efficiently, and to get on with life. They tolerate damage well and having their older outer leaves periodically removed by animals even benefits them.

A vivid illustration of the importance of having old outer leaves removed has come from New Zealand, an island group never reached by mammalian grazers until humans brought them. Extinct avian herbivores—above all, moas—fed by reaching in and plucking out leaves from the centres of tufts of grasses and sedges. To make up for the absence of grazing mammals that would have removed older, outer leaves from tussocks, most New Zealand pampas grasses (subfamily Danthonoidiaceae, Poaceae) actively shed their old outer leaves by abscission (Antonelli et al., 2011); they essentially bite off their own leaves. This is quite unlike the vast majority of the worlds' grasses that have co-evolved with mammalian herbivores and where abscission zones near the bases of leaves are absent.

Summary

Whereas this chapter has tried to emphasize the pressures exerted on leaves by folivores, the rest of the book turns the tables and looks at how leaves put pressure on leaf eaters. Several points related to this have already emerged. First, what animals want from leaves is nutritional resources for their own growth and reproduction. Plants need to conserve these same resources for themselves, so plant–herbivore interaction is a competition for energy and nutrients. One implication of this is that leaf defences have to be looked at in a social context. That is, when one species has leaves that are more nutritious than those of its neighbour it will need better defences. Seen from an energetics perspective, many of the strategies that leaves use to protect themselves make sense: many defences specifically reduce herbivore feeding efficiency. Related to this is a point needing frequent reiteration: leaf defences have co-evolved in remarkable ways with the digestive systems of herbivores. Additionally—and using a range of mechanisms—plants co-opt help from carnivores in order to reduce their own predation. Together, the leaf's direct and indirect defences help to explain why most plants that survive the seedling stage in nature live to near their full potential. Then there is the fact that leaves and their cells are attacked by a remarkably broad range of vertebrates and invertebrates, whose numbers and diversity are so great that defences specific to one attacker are rarer than are general mechanisms aimed at suites of herbivores. Finally, not all defences fit neatly into the categories of direct or indirect defence. Some strategies, like tolerance, may be equally important. Among yet other defence-related phenomena that are still rather poorly documented are those concerning leaf shape and colouring. These form the starting point of an investigation of how leaves defend themselves from herbivores.

2

Leaf colour patterning and leaf form

The visual appearance of a leaf, its shape and colour, is usually the first of its properties to confront the herbivore. Whereas leaf form is remarkably variable and often indicates adaptations to the physical environment, some leaf features have been selected in response to biotic pressures. When atypical leaf forms occur these are often associated with crypsis, a survival strategy which is well known and quite well documented and that can reduce herbivory. However, this chapter dwells principally on leaf colour. Many patterns on leaves, such as white or red spots, make plants more visible and therefore highly attractive to gardeners. But what if any roles do these markings serve in nature? Is their production likely to come with a cost for the plant? The starting point in an attempt to answer this is another question of long-standing interest: 'Why are leaves green?'

The palette of colour available to leaves

Chlorophyll makes leaves green, but why was this type of pigment selected and conserved in nature? After all, chlorophylls do not absorb all the light that falls on a leaf, and, although they absorb some wavelengths remarkably well, they do not do so across the full visual spectrum. Leaves absorb blue and red light efficiently, whereas much of the green light passes through the leaf or is reflected, therefore not all the light energy potentially available to leaves is exploited. So chlorophyll would, in terms of light gathering, appear to be suboptimal. In theory, black leaves would be better at capturing solar energy than green leaves, since the former could, in principle, harvest far more light. But there are no truly black forests. Instead, leaves limit the energy input that they receive and it has been suggested that being green reduces the extent of collateral damage that can result when photosystems in chloroplast membranes capture too much light energy (Terashima et al., 2009). This is only one way leaves deal with excess light and another mechanism works at a different level and involves dynamic changes in chloroplast positioning. The hue of the leaf in nature changes almost constantly (but imperceptibly) as light conditions change. In low light, chloroplasts migrate towards the upper surfaces of photosynthetic cells, but when too much light falls on the leaf they are aligned against cell walls that are parallel to incoming photons. This is a transient process; the rest of the chapter deals with more stable leaf patterning.

The ability to decorate the surfaces of otherwise plain green photosynthetic tissues appears to be ancient; for example, it occurs in some spikemosses (Selaginellaceae) which bear colour

Published 2014 by Oxford University Press.

patterns on their leaf-like microphylls. Patterned leaf surfaces are relatively common in angiosperms, but some colours are more common than others. In theory, plants can add any colour to their upper, light-receiving leaf surfaces provided that these colours do not impinge on the light absorbed by chlorophyll. However, the fact that chlorophylls absorb strongly at both the blue and the red side of the spectrum, and to a small extent between them, means that the range of colours that are usable for leaf decoration is limited. What are the consequences for the plant of producing coloured or even colourless patches of tissue on leaves?

Colouring leaves comes at a price. The 'cheaper' colours are dark, and plants including long-lived trees with red, purple, or even brown foliage can survive and grow well in gardens and parks, a good example being the copper beech (*Fagus sylvatica* var. purpurea, Fagaceae). In such plants, reddish or purple pigments (chiefly phenolics) are made in higher-than-normal levels and these absorb the green light that is not strongly absorbed by chlorophyll. So photosynthesis in the leaf can continue at only marginally reduced levels and purple-coloured leaves on trees such as copper beech do not interfere noticeably with growth. It is therefore likely that using certain dark pigments more sparingly on leaves, for example as spots or stripes, will have an even smaller impact on growth. This, indeed, is borne out by the fact that wild plants with dark spots on their leaves are relatively common. In contrast to red or purple, blue is a rare colour on leaves probably because blue pigments absorb red light. When blue is found on leaves it is usually created by physical interference, not by chemical pigmentation, as in the blue surfaces of some tropical *Selaginella* species such as *S. wildenowii* in Thailand and Malaysia. Orange on leaves is also rare, although it is used to spectacular effect on the young leaves of the croton *Codiaeum variegatum* (Euphorbiaceae) throughout the IndoPacific.

Another way to produce patterns on leaves is to produce leaf sectors lacking chlorophyll so that the tissue appears white. This is not only a potentially good way of making a strong pattern—it could, in theory, be used to make a white background on which to put pigments of any colour. Given the prevalence of certain variegated house plants and ornamentals it is easy to get the impression that plants with extensive white sectors on their leaves are common in nature. But white-variegated leaves are more the exception than the rule and they are particularly rare, for example, on the adult foliage of tree leaves. With this in mind it is usually hard to see how colour markings would confer physiological benefits to the leaf. The problem is that having white patches usually requires reducing chlorophyll levels or masking chlorophyll with other pigments. This is more costly (unless the plant is a holoparasite) than producing a pigment that absorbs light where chlorophyll does not. Instead, a solution to generating pale colours on leaves is usually to reduce chlorophyll without eliminating it completely. When this is done, new hues can be generated. Clearly, marking leaves with certain colours comes at a price that will only be paid if it promotes the plant's survival. This appears to be the case in some plants. What colour patterns can do for the leaf in terms of defence against folivores can be divided into two categories the first of which does not require the herbivore's eyesight. The second category plays on the herbivore's ability to see, either making the plant hard to find, confusing the herbivore, or even warning it that the plant would be dangerous to eat. Exposure colouration, however, falls into the first category.

Folivore exposure colouration

Exposure colouration exploits the obvious danger for small herbivores of being seen by predators. Birds, for example, have good colour vision; they can detect insects even if the contrast difference between the insect and the leaf is low. To counter this, many small herbivores are of a colour or hue that is similar to the majority of the leaves they feed on. But this gives plants the possibility of exposing predators to view and, probably as a protective strategy, many plants produce flushes of new leaves that are pink, red, pale green or even white (Lev-Yadun et al., 2004). This phenomenon is widespread and, in each case, only when the young leaves are expanded and start to become more fibrous do they turn dark green. Taking only one example among many from warm climates, there is cocoa (*Theobroma cacoa*, Malvaceae), a tree that produces red leaves that become green once they are fully expanded. A pink or red background, being the complementary colour of green, is the worst possible habitat for a green caterpillar to hide in. Pale green is a more frequently encountered colour for new leaves in northern temperate forests and few are as visible as the emerging needles of spruces such as the Norway spruce (*Picea abies*, Pinaceae; Figure 2.1). Any insect that strays on to these tender young needles is likely to be picked off by a bird. But how can we know if this phenomenon is likely to have evolved to protect young foliage from invertebrates rather than playing some other role, for example in the interaction of plants and vertebrate herbivores?

Figure 2.1 Putative exposure colouration in the young needles of Norway spruce (*Picea abies*).

If plants with young red, pink, white or pale green leaves or leaf venation only arose through interactions with vertebrate herbivores then we would not expect to see it in places where these animals have had little impact on the evolution of the vegetation. However, on such an island—New Caledonia—there are innumerable excellent examples of this phenomenon (mostly having pink or red young leaves) in the endemic flora. So it would therefore seem unlikely that exposure coloration evolved to counter vertebrate herbivores. Exposure colouration works even if a herbivore is blind, but this is not the case in the following examples.

Folivores are sensitive to leaf hue and patterning

Humans, the majority of whom have trichromic vision, probably do not see the world in the way that most vertebrate leaf eaters do. Most other mammals (including most monkeys) have, in addition to a single type of rod cell photoreceptor, two cone photoreceptors giving them dichromic vision. Ungulates, for example, probably experience a more limited range of colour than we do, being something like the equivalent of red–green colourblind (Birgersson et al., 2001; Kelber et al., 2003). By contrast, many birds and reptiles use four types of cone cell photoreceptors for distinguishing colours. All these vertebrates would be expected to be able to see colour and hue very well, as would most insects which have, at a minimum, dichromic vision. Herbivores depend heavily on good eyesight and this has been borne out by both observation and experimentation.

Deer appear to associate the high chlorophyll levels in healthy Norway spruce needles with a high nitrogen content. For example roe deer (*Capreolus capreolus*) in Sweden prefer to browse on vigourous, dark green spruce seedlings rather than lighter green or yellowish plantlets (Berquist and Örlander, 1998). This animal is sensitive to the hue of leaves and even though the dark green spruce seedlings are the most vigorous in the population (and therefore likely to be the best defended), they are the most sought-after. Based on this it would be reasonable to assume that anything that a healthy plant could do to make itself appear sickly might afford it some protection, at least until a herbivore learned that this was a deceptive colouration. Alternatively, to counter a herbivore's learning ability the plant could be polymorphic for leaf colour. Indeed, polymorphism in leaf shape and patterning are relatively common and are often associated with herbivory. In this case, a herbivore that, having tasted the leaf and found it good, might establish a search image and avoid otherwise similar forms that differ visually. This scenario was put to the test with sheep feeding on white clover (*Trifolium repens*, Fabaceae), a common plant in which the leaves often have a series of different white markings. Given the choice between plants with unmarked leaves and several morphs, each with different white marks on their leaves, sheep preferred the former (Cahn and Harper, 1976). Here, the ungulate distinguished the presence of a rather small pattern on healthy leaves and based its feeding preference on this. If this happens in nature then morphs with different patterns on their leaves would likely be under less pressure from the herbivore than the plants with unmarked leaves. The opposite scenario whereby a patterned morph was recognized by a herbivore that passed by a non-patterned morph would also work in favour of the plant.

Assessing the role of leaf patterning in the absence of field tests such as those performed by Cahn and Harper is difficult for many reasons, not least of which is that the colours on leaves or bracts of many plants have roles in attracting insect pollinators to flowers (e.g. the red bracts of *Poinsettia*). Nevertheless, many patterned leaves seem to be produced for reasons that are unrelated to the flowering process. Instead, colour patterns on numerous leaves are likely to protect these organs from herbivory. For convenience, patterns on leaves can be divided into two classes. Those that are non-deceptive and those that are deceptive. A non-deceptive pattern typically carries a legitimate warning.

Non-deceptive colouration: warning

Given the fact that herbivores can establish search images for certain leaves that they feed on, it is also possible that they might be able to associate colour patterns on leaves (or leaf-associated organs) as warnings. Such non-deceptive warning (aposematic) patterning is used frequently by animals—chiefly by small animals that bear warning signs exhibited to vertebrate predators. There is no reason to think that similar phenomena do not exist in plants. Warning colouration, if it exists on or near leaves, should be easily seen but would not necessarily require a change of overall leaf form—just as the warning patterning on the abdomens of hornets and wasps does not require a change in body form relative to bees with undecorated abdomens. In the case of plants, patterns on leaves could warn animals of the presence of toxins or they might draw attention to various physical defences such as thorns (Lev-Yadun, 2001). Sadly, however, there is little hard evidence for the use of warning colouration on leaves and this is a particularly challenging area in leaf biology (Wiens, 1978). Nevertheless, it is likely that the time of day that a herbivore feeds will be a key issue in the choice of warning colour.

Throughout the world there are countless examples of leaf eaters that are nocturnal or cathemeral, that is they feed in both periods of daylight and at night. This is as common in African or Indonesian ungulates (Estes, 1992; van Schaik and Griffiths, 1996) as it is for those in the forests of Europe and the Americas. Similarly, the majority of marsupials are active at night. All these animals have good night vision (Kelber et al., 2003). They should be able to see white patterns on green leaves in low light and to see tonal differences between what most humans see as green and red. All this is relevant, and any analysis of the roles of leaf hue, colour or patterning in plant-herbivore interaction can only have meaning if it is related to the times of day at which vertebrate folivores feed. The following section uses this premise in an attempt to illustrate what are still putative examples of plant warning patterns. For convenience, patterns are separated into those that are red (or purple) and those that are pale or white. Nature, of course, is more complicated and some leaves display both colour divisions.

Red

Poisonous plants with dark flecks on their stems, leaves or reproductive organs are quite abundant—an infamous example being hemlock (*Conium maculatum*; Apiaceae). In this case it is the stem that is patterned, not the leaves; but the principle is the same. Dark flecks may be warnings. Distinctive red or purple markings on fully expanded leaves (as opposed to entirely red leaves) are rare on trees and more common on herbs, often those growing in fairly bright sunlit conditions in forest clearings or among grass. Any folivore that avoids feeding in the full sunshine and is more active as the light fades might easily see white marking on green, but dark markings on a green background will to some extent fade out at night. Dark colours such as reds displayed on green leaves could, however, be seen clearly in daylight by herbivores as tonal changes.

What animals, if any, would be most likely to use such markings as feeding repellents? A clue to this comes from looking at the distribution of plants with large red flecks on their leaves. One frequently sees these plants where livestock are dominant herbivores and have been for centuries. Leaves with red markings often occur in Europe where there are domestic ungulates such as cattle. For example, the alpine dock (*Rumex alpinus*, Polygonaceae) often has large red areas on its leaves. This plant frequently grows in abundance next to cattle pens in mountainous regions. Outside Europe in central Asia there are many plants with red-flecked leaves, including various hyacinths and tulips as well as numerous orchids. Again, many of these are encountered where there are livestock. Could there be an association between diurnal feeding and the presence of plants with red-flecked leaves? If so, a good place to look for such plants would be where there are few wild, night-feeding herbivores but where many domestic ungulates feed during the day. Southern Ethiopia is one such place.

The Liban plateau region in south central Ethiopia has extensive grasslands that are heavily grazed by cows, goats and sheep all of which are corralled at night. On the most heavily grazed grassland, where only a few plants such as spiny acacias grow above the level of the closely cropped grasses, one can find *Kalanchoe marmorata* (Crassulaceae) with easily visible red flecks on its leaves (Figure 2.2). There is no doubt that these plants are accessible to the hundreds of cattle, sheep and goats that graze in their vicinity on an almost daily basis. But the little damage seen on the *Kalanchoe* leaves suggests that these plants are rarely even tasted, and one can at least offer the interpretation that these markings serve as warnings to day-feeding mammals.

The same might be true of the leaves of spotted orchids but here one is confronted with the issue of possible roles of leaf spotting in pollinator attraction. Fortunately, this can be resolved. For example, several *Cypripedium* species in East Asia (e.g. *C. forrestii* in southwestern China) can produce spectacular purple spots on leaves that are proximal to the short-stemmed flowers and many of these and large numbers of other orchids have large red or purple spots on the flowers themselves. Although it is possible if not likely that these patterns may themselves be warnings (flowers need both to attract pollinators and to exclude herbivores) it is necessary to look at simpler cases of leaf patterning that appear to be independent of flowering. Where can other examples of this type of patterning be found?

Some European orchids in the genus *Dactylorchis* display their leaf flecking most intensely on the youngest leaves and these spots often start to fade prior to flowering. The leaves of these plants have no spots on the lower surface, the flecks being positioned to be seen easily from above (Figure 2.3). Moreover, the leaves of spotted orchids in Europe are highly distasteful, leaving a bitter, soapy taste in the mouth: none are listed as being edible (Couplan, 2009).

Figure 2.2 Dark leaf flecks as putative warning signals. *Kalanchoe marmorata* (about 25 cm in height) growing on the Liban plateau, Ethiopia. (See Plate 3.)

Figure 2.3 Spotted leaves on orchids may serve as warnings. *Dactylorchis latifolia* growing in Fribourg, Switzerland. The dark purple leaf markings are displayed only on the upper leaf surface. They are present well before flowering is initiated and, upon the initiation of production of the floral stem, they often fade. Scale bar = 2 cm. (See Plate 4.)

Whereas red and purple may be employed as daytime warnings, white patterning is perhaps even more common on leaves. But its roles might be even more difficult to elicidate.

White

Some white-spotted plants (e.g. some aloes from Africa and Arabia) grow in the brightest of conditions but perhaps the majority of plants with white-spotted leaves do not. Instead, they are more usually found in broken light or shade— often on the floors of dense forests. Being white offers the best chances of being seen in low light. It is in the shade or in the night that colour differences fade and pale elements in the vegetation stand out. Sometimes this can be particularly striking as it is in African acacias (*Vachellia* spp., formerly *Acacia*), the vast majority of which have white thorns. When these plants dominate plant communities, as they often do in many of the drier regions of Africa, they can change the colour of whole landscapes. In the case of the whistling thorn acacia *Vachellia drepanolobium* (Fabaceae) of northeastern Africa (Figure 2.4A) this can give the scenery a silver-white hue that is visible from far off. Not always as visible at a distance are the gazelles and other ungulates that feed on the leaves of these small trees, most of them doing so well after sunset. One can speculate that the white thorn acacias found in Africa and elsewhere serve to warn vertebrate folivores that the leaves they feed on are well protected. They are, perhaps, seen by the herbivore as an indication of the plant's defensive vigour.

White or pale green warnings could, like their red counterparts, indicate the presence of powerful chemical defences. They might therefore be expected to occur in some latex-producing plant families. Consistent with this, some of the most impressive white markings that I have seen in the wild were on the stems of *Euphorbia cactus* (Euphorbiaceae) growing in Saudi Arabia (Figure 2.4B). Although not on leaves, the fact that these stem markings were highly visible and invariably aligned with physical defences (thorns) underlines a likely role in warning that the plant is extremely well defended not only by the thorns themselves but also with an acrid chemical-rich latex. So white patterning may have diverse roles to play on leaves, and one of them might be to serve as a warning to herbivores that feed in low light. However, two other possibilities complicate the interpretation of white markings on leaves. The first is the possibility that plants use them as a form of camouflage.

In some cases, spots on leaves may have evolved to mimic the environment. Work on this has been conducted on forest plants in the north eastern USA and this has revealed an interesting correlation between the presence evergreen trees in forests and mottling (mostly white) on leaves of smaller plants. Nine out of 12 species with mottled leaves grew in forests with evergreen trees; only three species with mottled leaves occurred in deciduous forests (Givnish, 1990). Furthermore, most of the mottled plants had leaves in early spring or even throughout the winter, not only during the summer. The idea is that these leaves are hard for vertebrates to see in environments where sunlight is broken, where the forest floor is illuminated with dappled light over an extended period of the year. Ideally, in cases like these it would be necessary to test the camouflage hypotheses by exclusion of mammals. But, in the absence of such tests the problem has to be approached more theoretically. In doing so it is

Figure 2.4 White as a putative warning signal. (A) *Vachellia* (*Acacia*) *drepanolobium* near Mega, southern Ethiopia. This species, like the majority of African acacias, has white thorns. (B) *Euphorbia cactus*, Jebel Fayfa, southwestern Saudi Arabia. Note that the white stripes correlate with the position of thorns. Scale bars = 1 cm.

tempting to look at how we humans react to mottled plants and why we readily choose them as garden ornamentals.

It is notable that we choose many white-spotted plants as garden plants and they are selected for their striking foliage, not because they are difficult to see in broken light. While the spotting on the leaves of many ornamentals does indeed resemble sun flecks, I do not feel that the camouflage hypothesis is compelling because the human brain quickly establishes search patterns for particular species of plant with patterned leaves. Presumably other animals do too. Where the camouflage hypothesis is more convincing it is when it is combined with altered growth forms.

Deceptive colouration: colonization and damage mimicry

There are further ways in which leaf patterning may afford protection from herbivory. One of these, which is unrelated to camouflage, would be to deceive herbivores into thinking that an otherwise healthy leaf was either a low quality habitat or a low quality food. This would amount to quality reduction mimicry and would suggest to the herbivore that the leaf was either colonized, damaged, or of low quality. In the case of colonization mimicry, leaves can display shapes that resemble insect eggs or that suggest that they are home to leaf miners. Alternatively, they produce spots or squiggles on their leaf surfaces, suggesting that the leaves may be damaged or soiled. Most of these types of mimicry make leaves appear to have herbivores in them or on them, or they might otherwise make the leaf seem unattractive. Fortunately, this area, unlike the less researched area of warning colouration, has been the subject of much observation and even some experimentation. Colonization mimicry appears to be widespread and mostly directed at insects. As a defence this can only deter attack; once the insect starts feeding or egg-laying it is probably too late for it to be effective. Some of the most celebrated examples of this type of mimicry involve passion flowers (*Passiflora* spp.; Passifloraceae).

From a defence point of view, *Passiflora* is a creative genus that displays a broad sweep of defence strategies that range from making cyanogenic glycosides to using extrafloral nectaries to attract protective ants. But the leaves of some passion flower species go even further, using visual cues that deceive female heliconid (*Heliconius* spp.) butterflies into perceiving leaves as being already egg-laden (Gilbert, 1975). That is, a few *Passiflora* species produce pale yellow or white structures that look like the eggs of these butterflies. One of the reasons that this is thought to work well is that heliconid caterpillars often eat each other, so laying too many eggs on one leaf must be avoided. The leaves of *P. boenderi* show variegation along two of their three main leaf veins, and near these are bold white egg-sized patches on the upper leaf surface (Figure 2.5). Although not described as such in the original publication (MacDougal, 2003), the undersides of most of these egg mimic patches are extrafloral nectaries that the plant can use to attract protective ants. So the same area of leaf appears to have two different defence-related roles to play—one for each leaf surface.

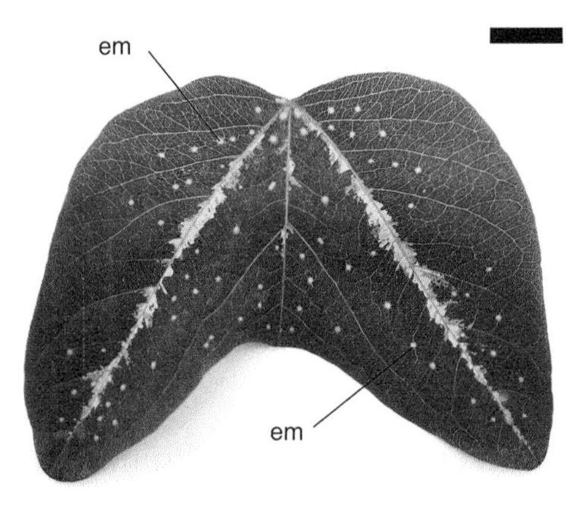

Figure 2.5 Deceptive patterns on passionflower leaves. To deter egg-laying by heliconid butterflies the leaves of *Passiflora boenderi* from Costa Rica have yellow egg mimic (em) spots on their upper surface. Not shown is the underside of the leaf. Below the egg mimic spots are extrafloral nectaries that produce sugar to attract protective ants. (See Plate 5.)

There are other examples of foliage patterning in the New World that are probably directed at stopping leaf miners laying their eggs in leaves. One report of this concerns a climbing plant from rain forests in the Andes (*Byttneria aculeata*, Sterculiaceae) where individuals with variegated leaves were found to be less infected by miners than those with plain green leaves. The interpretation of this was that the variegations on leaves mimicked leaf miner tracks and that this reduced egg laying (Smith, 1986). This example is not alone. The genus *Caladium* (Araceae) provides some of the really eye-catching examples of leaf decoration. Among these is *C. steudneriifolium* in southern Ecuador, where this species is polymorphic for white markings on its juvenile leaves. Those individuals of the same species with plain green leaves displayed more leaf miner damage than plants with white-patched leaves. Moreover, when white markings were painted on to plain green leaves, damage levels fell about 20-fold relative to the control plants (Soltau et al., 2009). One has to ask how much of the sometimes spectacular patterning and colouration observed in plants such as *Begonia* (Begoniaceae) or in the Marantaceae (e.g. in some *Calathea* and *Maranta*) serves these functions.

Throughout the world one can find countless other examples of what might be damage or colonization mimicry based on leaf mottling. But the word 'might' is important. Leaf mottling, instead of being a dishonest ploy to make a herbivore think a leaf has already been attacked, could be an honest warning that the leaf is toxic or otherwise well protected, or it could be both. A common plant from Europe can be used to illustrate the difficulty of interpreting leaf patterning in the absence of experimental support.

Finding a clump of the common lungwort (*Pulmonaria officinalis*, Boraginaceae) growing in the dark shade can be spectacular. Under these conditions, most have striking irregular spots on the leaves (linking this patterning to light intensity or quality, these spots intensify if the plants grow in dark shade). So finding these plants hidden in the undergrowth in northern Europe can be as rewarding as glimpsing a spotted *Caladium* in a Central American forest. But why the lungwort's leaves are spotted is unclear. The example in Figure 2.6 shows a lungwort leaf with large irregular flecks that are mostly found near the leaf centre. These large markings may closely resemble bird droppings. On the same leaves are smaller spots which are more concentrated at the lamina edges. The plant in Figure 2.6 was growing in the presence of horses and was associated with two other plants that have good antivertebrate defences: common stinging nettles and hemlock. Are the spots on the common lungwort acting as warnings or are they damage mimicry? Could they even suggest to large herbivores that the leaf has been soiled by birds? Given the associated plants, the fact that the Boraginaceae often has good antivertebrate defences, and the fact that the leaf surfaces are covered with brittle hairs, I would speculate that these leaves are patterned to be quickly recognized and avoided by mammals. This would not exclude similar roles in defence against invertebrates.

In summary, determining exactly how leaf decoration works in protection is complicated by the possibility that patterns may serve multiple functions, acting, for example, as deceptive egg- or damage-mimics to deter invertebrates and at the same time perhaps acting as honest warnings for vertebrate herbivores. But the big picture should not be lost because of

Figure 2.6 Enigmatic white patterning on a lungwort. Common lungwort (*Pulmonaria officinalis*) growing in southern England in the shade of nettles. Scale bar = 1 cm.

this. Both of these mechanisms would contribute to leaf defence—as would exposure colouration. In the absence of good evidence that they contribute useful physiological functions, leaf patterns probably all boil down to the same thing—protecting plants from animals. It is now time to use the power of molecular genetics to manipulate leaf patterning and test its effects on herbivores under controlled conditions.

Dissimulation and crypsis

Mimicry of soil, sticks, and stones

Plants can hide very effectively in their environment and there is more evidence that they do this than there is for the use of leaf patterns as warnings. Some species resemble the colour of the ground they grow on, or can look like dead branches or stones. Others mimic more common plants or those that dominate their habitats. The use of crypsis strategies starts early in life where germination and emergence into the light immediately puts a seedling into the visual domain. To avoid being found quickly, one strategy is to partially mask the colour of chlorophyll with darker pigments either painted evenly over the cotyledons or displayed as flecks. A striking example of the latter is seen in the Mount Teide bugloss (*Echium wildpretii*, Boraginaceae) from the Canary Islands. The cotyledons of these seedlings often have soil-coloured line markings, presumably to help them blend into the terrain on which they grow. Mimicry of soil colours in the adult phase is common and frequently observed in dry environments in a wide variety of plant families. Often, this form of mimicry can be impressive because the plant has a colour that closely resembles the dominant soil or stone type, even when its surface has been washed. Many examples of this occur in the New World, Africa and

Arabia, with an abundance of examples concentrated in northwestern South Africa and southwestern Namibia (Namaqualand) in many plant families that are sought-after by collectors of succulents, e.g. Azioaceae, Apocynaceae, Crassulaceae, Euphorbiaceae, and Portulacaceae.

When crypsis strategies are employed by plants this often involves changes in form whereby some members of a genus break from the norm and develop to resemble members of other genera or even other families. Since phasmids (leaf and stick insects) can mimic dead branches or damaged leaves it is presumably at least as easy for plants to evolve this type of strategy and it is indeed found in many parts of the world. There are, for example, leaves with highly irregular edges that look as though they have been damaged by insects, e.g. *Ficus soroceroides* (Moraceae) from Madagascar. But there has been little published research on whether or not such leaf indentations serve the plant by reducing herbivory. Dead branch mimicry can work so well that plants doing this are rarely noticed and people seeking them have to consciously exclude anything that looks alive. Members of the genus *Rhytidocaulon* (Apocynaceae) on the Arabian peninsula and the Horn of Africa resemble dead wood so well that experienced botanists looking for these plants usually have to wait until they are in flower. Additionally, *Rhytidocaulon* and related stick-resembling *Caralluma* species often grow sprawled on the ground below other vegetation and this adds to the effort required to find them.

Then there is the better-known phenomenon of stone mimicry as exemplified by the Aizoaceae family in South Africa. This family has produced scores of stone mimics but it is not alone. Other families where the same phenomenon has occurred in multiple genera include the Apocynaceae and the Liliaceae, with the latter family showing that this phenomenon is by no means restricted to dicotyledons (Wiens, 1978). Dead branch mimics and stone mimics are particularly frequent in warm open habitats where they presumably reduce the chances of plants being discovered by herbivores—particularly those that are low to the ground. It is notable that such plants tend to be most common where reptilian herbivores exist.

Mimicry of grasses and other plants

Grasses dominate vast land surfaces and are so successful that they have been copied. This strategy, especially at the seedling stage, is fairly widespread and may serve to hide a nutritionally valuable plant in a crowd. But whereas many seedlings vaguely resemble grasses (many seedlings in the Apiaceae have the propensity to do this), it is hard to tell whether this really is mimicry of another plant. A first necessity is to observe the plants in their native environments. It is in fields of native grasses where the mimicry of a grass will work best. In Australia, for example, the seedlings of grass trees (*Xanthorrhoea* spp., Xanthorrhoeaceae) can be hard to find among true grasses. But the phenomenon of grass mimicry is even more spectacular in some spear grasses (*Aciphylla* spp.; Apiaceae) in New Zealand where the seedlings closely mimic the dominant grass species among which they grow.

The New Zealand botanist David Glenny and I visited fields outside the city of Lincoln on the South Island to look at this. We found fields that were dominated by the endemic grass *Festuca novae-zelandiae* among which were adult plants of the spear grass *A. subflabellata*. It is the young of this plant that are difficult to find. So in the search for spear grass seedlings turf samples were excavated and examined carefully. This process was time-consuming, and it was only when all plants in the turf were separated that we could distinguish the seedlings of *A. subflabellata* from those of the *Festuca*. In this case there is a good hypothesis to explain how this evolved. It would have helped the spear grasses to escape predation by now extinct birds (moas), the only large vertebrates on New Zealand capable of submitting the flora to the kinds of pressures that are likely to lead to the evolution of mimicry (Atkinson and Greenwood, 1989). And there are further interesting examples when it comes to mimicking grasses. In the Southwestern USA the grama grass cactus (*Pediocactus papyracanthus*) mimics the dead leaves of a dominant or co-dominant grass, blue grama (*Bouteloua gracilis*) (Wiens, 1978). Many dicot seedlings mimic grasses; the opposite appears to be rarer.

Still in the realm of interplant mimicry there are examples where it appears that highly toxic dicotyledonous plants are mimicked by less toxic plants from other plant families. *Tylecodon wallichii* (Crassulaceae) is a toxic succulent that appears to be mimicked by *Othonna herrei* (Asteraceae) in Namaqualand in southwestern Africa. Far away from these plants, in New Zealand, the bitter leaves of the mountain horopito (*Pseudowintera colorata*, Winteraceae) can have vivid purple spots on their pale leaves, which are visually similar to the less bitter leaves of *Alseuosmia pusilla* (Alseuosmiaceae). It is possible that these are cases of Batesian mimicry in the plant kingdom.

Mistletoe host mimicry

Some of the most stunning examples of interplant mimicry are found in several genera of mistletoes from the Loranthaceae in Australia. To see this, an authority on mistletoes, the Australian botanist Bryan Barlow, took me to Tidbinbilla near Canberra to look at infected casuarinas and eucalyptus trees. Casuarina trees (*Casuarina* spp., Casuarinaceae) are angiosperms, but their growth form more closely resembles that of many conifers—they have long, needle-like photosynthetic shoots with the leaves reduced to small bracts. We stopped to look at a large casuarina that appeared at first sight to be perfectly healthy—but the tree was in fact very heavily infected with the mistletoe *Amyema cambagei*. Remarkably, this parasite's own leaves were visually similar to the casuarina shoots, only being easily distinguishable to me from close up where, side-by-side (Figure 2.7), the mistletoe leaves were broader and slightly lighter green than those of its host. Even with practice it can be difficult to spot casuarina mistletoes unless they are flowering, and even then *A. cambagei* is quite discrete, having less showy flowers than those of many of its relatives.

The mistletoes of Australia do not restrict themselves to parasitizing casuarinas, and most of the dominant tree genera on this continent (*Acacia, Eucalyptus, Melaleuca, Rhizophora,* etc.) have at least one of these parasites that mimics their phyllodes or leaves (Barlow, 1981). In most cases, and as expected, the mistletoe mimicks the adult-phase foliage of its host,

Figure 2.7 Host leaf mimicry by a mistletoe. Leaves of a *Casuarina* tree on the left are compared with those of the mistletoe *Amyema cambagei*. Tidbinbilla, near Canberra, Australia.

although *Eucalyptus shirleyi* retains its juvenile leaves. This tree is parasitized by a mistletoe (*Dendrophthoe homoplastica*) that closely mimics the juvenile host leaf form (Barlow, 1981).

What selective advantage, if any, does the parasite obtain from resembling its host? Barlow's hypothesis for this is compelling. This mimicry is thought to come into its own at night when marsupial folivores feed. It is likely that tree-climbing animals, mostly possums, have trouble finding mistletoe foliage among host leaves and that they provide the selection pressure favouring the evolution of this remarkable mimicry. Barlow provided a convincing argument that two marsupials in particular, ringtail possums (*Pseudocheirus peregrinus*) and brushtail possums (*Trichosurus vulpecula*), exert much of this pressure. These herbivores clearly eat mistletoe leaves, and in many tests preferred them to the leaves of the host trees. Everything seems to fit well with this mimicry being directed against marsupials, but the possibility that it evolved to stop invertebrates finding the mistletoes also has to be considered (Barlow and Wiens, 1977). While this is not to be excluded, it seems less likely when one compares Southern and Northern mistletoes. Cryptic mimicry among mistletoes is not a purely Australian phenomenon and some other mistletoes in other Southern floras have a tendency to hide in their hosts (Barlow, 1981). However, the widespread mistletoe *Viscum album* (Viscaceae) found in the northern hemisphere is, in comparison to cryptic mistletoes, relatively easy to spot on any of its multiple hosts. Arboreal vertebrate folivores are almost invariably absent where this mistletoe grows.

Crypsis through movement

Before moving on to other aspects concerning leaf surfaces it is necessary to mention a final trick that a few plants use to make themselves hard to find: leaf movement. Most leaf movements are slow, like the night (nyctinastic) movements of plants such as common wood-sorrel (*Oxalis acetosella*, Oxalidaceae) in Europe. This phenomenon can be extraordinary to witness in some species of *Marantochloa* (Marantaceae) in tropical forests of Equatorial western Africa. These plants position their leaves horizontally to the ground during the day but, depending on the species, either raise or lower them to vertical at night. But this is likely to be a mechanism to remove debris that falls constantly from the forest canopies and it also reduces the leaf surface that can be hit and damaged by dead branches, fruit and other falling objects. There is little reason to associate these types of leaf movements with defence against herbivores. In order to find examples of plants that employ a nyctinasty-related mechanism in defence it would be wise to start the search with families that are already known for their use of a particularly wide variety of defence strategies. This is the case for the Fabaceae, a large family with several touch-sensitive species including *Mimosa pudica*, a plant that, in additional to spectacular night closure, also displays rapid touch- and damage-induced leaf movements (Figure 2.8).

Mimosa pudica is a very successful weed. It has escaped its home in the warm and humid regions of the Americas to be encountered in many parts of the tropics. Just how strong and well-defined the movement responses of *Mimosa pudica* are depends on its vigour, that is, on its age and physiological status, and on the weather. In nature it seems to work best when it is hot and humid. Under these conditions, even a small insect can provoke a closure response that is restricted to the pinnae whereas rougher handling provokes the generation of signals that travel through the woody tissues; the rachis, the petiole and, in very severe wounding, even through the stem. When a leaf is bitten or crushed this signal is transmitted to the base

Figure 2.8 Crypsis through movement. The fast, touch-induced movements of *Mimosa pudica* reduce the visibility of leaves to vertebrate herbivores. For scale, the flower diameters are ~1 cm.

of leaves which then drop downwards as cells in the leaf base pulvinus collapse and the greenery of the plant disappears from view. In the words of Jagadis Chandra Bose (1858–1937), who devoted much of his research to understanding the movement-associated electrophysiology of this plant, 'Nothing could be more striking than the rapid change by which a patch of vivid green becomes transformed into thin lines of dull grey unnoticed against the dark ground. It is probable that this invisibility may serve as a means of protection' (Bose, 1926). No-one has come up with a more credible explanation for this behaviour.

3

Structural defences and specialized defence cells

Leaf surface defence cells and leaf surface habitats

If it has not been confused by crypsis or warned off eating leaves by colour patterning, the herbivore that has decided to feed will inevitably encounter the cuticle, a complex chemical structure secreted by epidermal cells that functions in limiting water and ion loss from the plant. In addition to these roles, the cuticle is a barrier to pathogens and might equally make it difficult for epiphylls such as algae and bryophytes to establish themselves, a constant hazard for a long-lived leaf in a humid forest. In some cases the cuticle can also prevent small animals climbing on to leaves. For example, ants that are symbiotic on *Macaranga pruinosa* (Euphorbiaceae) in southeast Asia can climb the vertical stems of this plant, but the stems are so slippery that non-adapted ants cannot rob sugar from the plant's extrafloral nectaries (Federle et al., 1997). Some of the chemicals responsible for making such super-slippery surfaces have been identified, for example in the castor bean (*Ricinus communis*, Euphorbiaceae), where some populations segregate for white-stemmed variants. It is the outermost cuticular layer of the stems of these plants that is coated with wax crystals made of pentacyclic triterpenes (Guhling et al., 2006). Why, then, are slippery cuticles not more common?

Tiny invertebrate herbivores, phytophagous mites for example, may spend much or all of their lives on a single leaf, so the nature of their habitat can have a profound effect on their chances of survival. But leaf surface topography can differ enormously between plant species and even between leaf surfaces on the same plant. What is the optimal leaf surface when it comes to defence—smooth and shiny, covered in trichomes and sticky, or somewhere in between? Smooth-surfaced leaves have their advantages. First, herbivores are visually exposed. Second—and of great importance during egg development—herbivores may risk the danger of desiccation since smooth leaf surfaces can be dry, hostile environments. However, from the plant's point of view, a disadvantage of a smooth, featureless epidermis is that it offers no place for predatory enemies of the herbivore to hide. By contrast, complex surfaces with many trichomes offer hiding places for the herbivores predators but also for herbivores. When it comes to defence there is no such thing as a perfect leaf surface.

The epidermis

Below the cuticle, the epidermis represents the first contiguous cellular barrier that is encountered by the herbivore. Although the shape of epidermal pavement cells differs between species, they commonly display obvious defence features such as highly reinforced cell walls and adaptations for the storage of chemicals, not all of which have primary functions in defence (some act as filters for ultraviolet light, etc.). The epidermal pavement is a formidable barrier for small arthropods. For example, most piercing insects such as aphids attack leaves by puncturing between or even through epidermal cells, and this is itself a difficult task since the pavement cells are often joined together in a jigsaw-like fashion, making them difficult to prise apart. An easier alternative would seem, at first sight, to be to use stomata or hydathodes as ports of entry.

Stomata and hydathodes

Stomata and hydathodes are potential weak points in the epidermis. They would appear to offer an ideal route for piercing-sucking insects to bypass defences in the epidermis. Yet this is probably rare. One reason is related to the positioning of stomata which are often placed away from vascular strands—the targets of insects such as aphids. Another reason may be that the substomatal cavity is defended physically and/or chemically by aptly named stomatal guard cells. Furthermore, stomata are dynamic and are touch- and damage -sensitive, closing in response to herbivore attack. Additionally, guard cells often have modifications suggesting roles in defence. This is exemplified in the stomata of *Equisetum* (horsetail) (Figure 3.1). What makes these stomata visually impressive is their silica teeth, but these are rather exceptional and more typical stomata may at first sight look like tiny undefended mouths. They are not: they are chemically armed. For example, the most abundant protein in the guard cells in *Arabidopsis* is myrosinase, an enzyme that hydrolyses glucose off a family of sulphur-rich defence compounds known as glucosinolates to generate toxic compounds (Husebye et al., 2002; Zhao et al., 2008). Not only are glucosinolate breakdown products poisonous, they also smell bad. A small herbivore getting too close to the stomatal mouth might be warned off by bad breath.

Trichomes: leaf surface defence

Trichomes—specialized cells or groups of cells protruding from the aerial surfaces of plants—originate from the epidermal pavement and have important roles in protecting leaves. They are common in virtually all angiosperm families and they range from simple, single-celled structures as in *Arabidopsis thaliana* to highly complex multicellular glandular trichomes found in plants such as wormwoods (*Artemisia* spp., Asteraceae) or in mints (*Mentha* spp., Lamiaceae) (Wagner, 1991). It is the chemicals (mostly terpenes) in glandular trichomes on the leaf surfaces of many of the Mediterranean spice plants that give them their agreeable smells, and, of course, contribute to defence. This trichome type typically contains several cell types including stalk cells, the biosynthetic cells that make the bulk of the chemicals in the trichome, and chemical storage cells (the glandular part of the trichome). It is also these

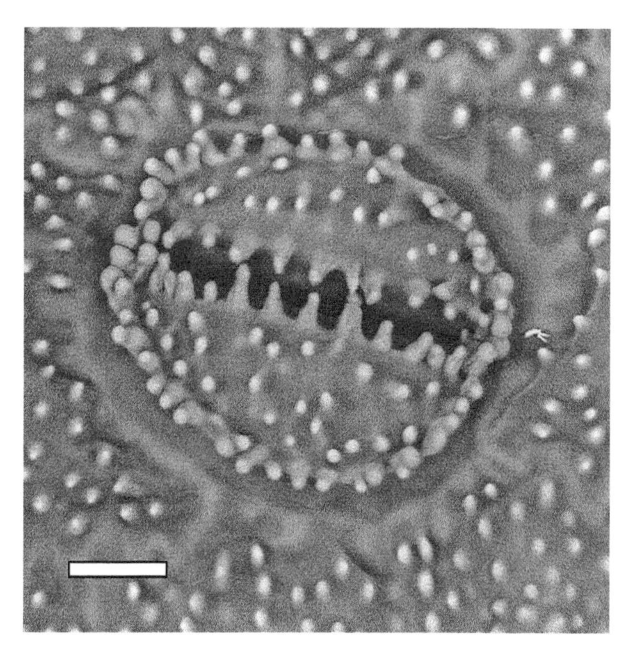

Figure 3.1 Stomatal defence in a horsetail. Stomata are often highly defended by chemicals. In the case of the horsetail *Equisetum arvense* from central Europe the guard cells have prominent teeth-like silica reinforcements that may help to protect the substomatal cavity from small herbivores. Scale bar = 10 μm. Image: A. Muccioli, Lausanne Electron Microscopy Facility.

sorts of trichomes that cover the leaves of tobacco (*Nicotiana tabacum*, Solanaceae), contributing almost 5% of the leaf weight and being where as much as 30% of phenolic chemicals in the whole leaf are found. But in many species the defences conferred by trichomes are largely or purely physical. Some, for example, act as hooks to catch the legs of invertebrates, hindering their movement. Spectacular examples of this are found in the Loasaceae (Eisner, 2003), a New World family.

Hooks are not the only way to trap tiny herbivores, and this can also be effected if the epidermis produces glues. When this occurs it is mostly the result of secretion from glandular trichomes or from closely related secretory emergences known as colleters. Sticky surfaces on plants are more commonly used for their protection rather than for trapping a food source (as in many carnivorous plants) and examples are widespread globally. The sticky geranium (*Geranium viscosissimum*, Geraniaceae) of the western USA is a good example, as is the great laurel (*Rhododendron maximum*, Ericaceae) of the northeastern USA that can trap quite large insects on its stems, petioles, and the abaxial midveins. Stickiness is most problematic for small unadapted invertebrates, and any insect caught on a sticky leaf surface is in grave danger of predation. This is exemplified by the shrub *Trichogoniopsis adenantha* (Asteraceae) in southeastern Brasil. It traps midges and ants on glandular trichomes on its leaves and stems and these are preyed on by lynx spiders (*Peucetia* spp.) which have cleverly

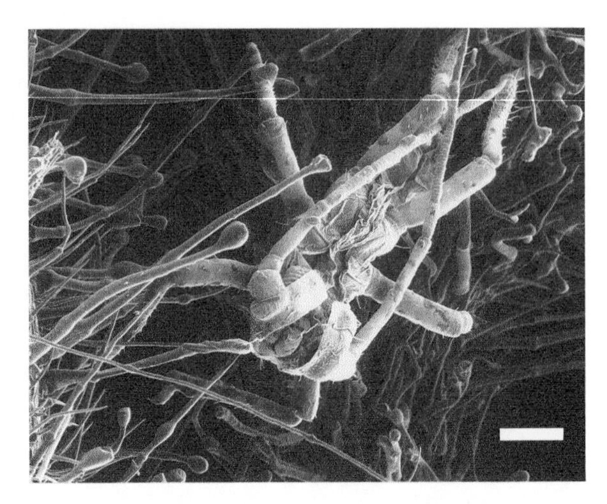

Figure 3.2 Trichomes as insect traps. The leaf and stem trichomes of *Cerastium glomeratum* have sticky tips that trap a variety of non-adapted invertebrates such as the insect shown in the image. Scale bar = 100 μm. Image: A. Muccioli, Lausanne Electron Microscopy Facility.

learned to associate themselves with these sticky-surfaced plants (Romero et al., 2008). Similarly, when dead fruit flies were placed on glandular trichomes on tarweed (*Madia elegans*; Asteraceae) from the southeatern USA, they attracted a suite of predators including lynx spiders (Krimmel and Pearse, 2012). There may be examples of this in Europe too. Invertebrates get caught on the stems and leaves of the stinking groundsel (*Senecio viscosus*, Asteraceae) and on sticky mouse-ear chickweed (*Cerastium glomeratum*, Caryophyllaceae; Figure 3.2), where, upon close inspection, insects body parts are often found. Bodyless heads are not uncommon. But if trichomes trap insects too efficiently they can also trap valuable predators in addition to herbivores (Eisner et al., 1998).

Evidence for the importance of trichomes in defence

Most plants that usually produce trichomes can grow well without them, and cultivating a plant in protected conditions under glass can result in leaves that, while similar in shape to those grown outdoors, are relatively denuded of these structures. Moreover, many trichomeless mutant plants exist and these generally grow similarly to wild-type plants allowing them to be used in experiments designed to assess the contribution of trichomes to defence. Such experiments have been performed with mutants of *Arabidopsis thaliana* and the results obtained by different laboratories agree well. For example, the larvae of generalist lepidopteran herbivores grew faster on trichomeless *Arabidopsis* than on the trichome-bearing plants, demonstrating a protective role of leaf trichomes against these insect herbivores (e.g. Dalin et al., 2008). But is this the full story? It is unlikely that anyone working in this field would be surprised if the genetic removal of trichomes in some plants had other effects on the plant's survival. But at least some roles in defence against folivores are now established solidly.

Despite being so accessible to study, many mysteries remain concerning trichomes; one of these relates to the fact that trichomes are often aligned or tilted in various ways. For example, the petiolar and midrib trichomes in *Cannabis sativa* are tilted towards the base of the leaf, and in *Arabidopsis* the beautiful trident-like leaf trichomes are almost all aligned with respect to the long axis of the leaf. Then there is positioning relative to the vasculature to consider. This is illustrated for the common nettle (Figure 3.3A); just one of the many defence features of these leaves is the strategic positioning of the small glandular trichomes. These are spaced regularly along the vasculature to protect it from incursions by invertebrates. Finally, trichomes may act as mechanosensors that immediately and autonomously respond to intrusion (Peiffer et al., 2009). An exciting prospect is to harness the power of genetics to

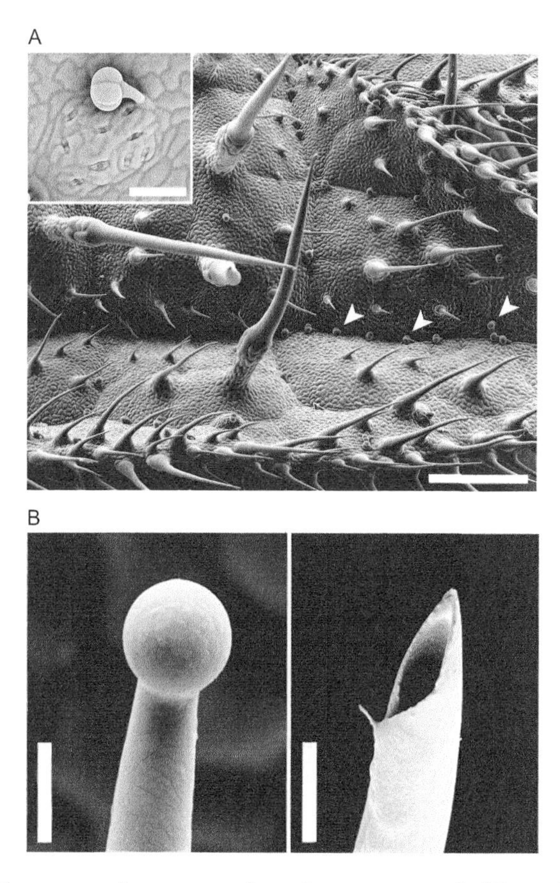

Figure 3.3 Defence features on the upper surface of a common nettle (*Urtica dioica*) leaf. (A) Stings dominate the image but glandular trichomes (indicated with arrowheads) are seen over the vasculature. Scale bar = 500 μm. The insert shows a single glandular trichome associated with a cluster of laminar hydathodes. Inset scale bar = 20 μm. (B) The tips of stings of *Urtica dioica*. The sting on the left is intact whereas the sting in the right panel is broken, showing the syringe-like point. Scale bar = 20 μm. Image: A. Muccioli, Lausanne Electron Microscopy Facility.

investigate the mechanism of mechanosensitivity in trichomes and to test the relevance of this aspect in defence. The same can be said for larger leaf surface defence structures, those that originate from more than one cell layer. Stings are good examples of such defences.

Stings

Stings are relatively rare in the plant kingdom and are distributed unevenly around the globe; the tropical Americas have more than their fair share of them. They are unknown in lycophytes, ferns and gymnosperms and, out of more than 400 families of flowering plants, only four families have members with stings. Within this scattered group, only three of the four sting-producing families, Euphorbiaceae (worldwide except the northern-most latitudes), Urticaceae (worldwide) and Loasaceae (mostly in the Americas south of Canada), contain multiple sting-carrying genera (Thurston and Lersten, 1969). The fourth family, Boraginaceae, is reported to contain stinging plants in only a single genus, *Wigandia* (e.g. *W. urens*, found in parts of the warm Americas) (Thurston, 1974), although trichomes on the leaves of many other members of this family can cause a sting-like sensation when touched. Being derived from both epidermal and subepidermal layers (and therefore being defined as emergences), sting structures differ somewhat between families, although the part of the sting that delivers the irritant is invariably made of a single large, elongated and living syringe cell with a rigid tip. In some cases it is known that the walls of this cell are silicified, making it particularly brittle so that it is easily broken off. This is the case in the widespread common stinging nettle (*Urtica dioica*, Urticaceae), where the often bulbous end of the syringe cell breaks off tangentially to its long axis to form a sharp point closely resembling the tip of a hypodermic needle (Thurston and Lersten, 1969; Thurston, 1974) (Figure 3.3B).

Stings have evolved to puncture epidermal cells in herbivores and to deliver, by injection, a fast-acting 'shock' defence barrier. The soft parts of herbivores are targeted, and touching stings by hand can only give a rough idea of how painful they might be to an eye or a lip. To experience this at first hand, one would have to try to bite stinging leaves straight off the plant and if this were done it is probable that the number of plants that we judge as being able to sting would increase (especially among the Boraginaceae). Moreover, whereas facial contact would be unpleasant with the nettles of the northern hemisphere, it would be unthinkable with some of their southern hemisphere counterparts. Luckily for humans, most plant stings are perceived immediately and fade within an hour or two, but there are some notable exceptions to this. One such plant is the Gympie nettle (*Dendrocnide moroides*) (Figure 3.4A) from northeastern Australia. These inoffensive-looking plants have long-lasting, debilitating stings where the pain sensation is said to return each time the affected skin changes temperature.

A frequent occurrence is that stings from the same species of plant can be quite variable in intensity and this can depend on both environmental conditions as well as on genetics. The petiolar stings of wild chayas (*Cnidoscolus* spp., Euphorbiaceae) in the inland forests of the southwestern Yucátan peninsula are unspectacular, in terms of the pain they inflict; whereas their counterparts in the brighter, drier north east of the peninsula often bear more

Figure 3.4 Stinging nettles from the southern hemisphere. (A) Leaf of the Gympie nettle *Dendrocnide moroides* from Queensland, Australia, that is known for causing long-duration pain. Scale bar = 1 cm. (B) Ongaonga (*Urtica ferox*) from New Zealand has a hard-hitting, fast-acting sting. Scale bar = 2 cm. (C) *U. ferox* showing detail of sting. Scale bar = 8 mm. (See Plate 6.)

painful stings. Similarly, it is often said that common nettles growing in shade have less violent stings than those of the same nettles growing in brightly lit habitats. Stronger, genetically based polymorphism has been found in *Urtica dioica* from southeastern England where some populations of this nettle were almost completely devoid of stings on their leaves (Pollard and Briggs, 1984a). But there does not seem to be a simple all-encompassing relationship between (i) sting density and intensity and (ii) plant habitat. Stinging plants are found in arid regions, humid tropical forests and even on mountains. What is clear is that stings, like thorns, can be an indication that a plant is of high nutritional quality relative to its neighbours—stinging plants are cooked in many parts of the world.

During the sting process the needle tip itself can break off in the skin where it may have a relatively long half-life due to its high content of enzyme-resistant cell wall components. These fragments are irritants, but they are masked, at least in the short term, by the pain-inducing sting fluid. Some plants have syringe cells that also contain quite spectacular crystalline deposits, as is the case for at least some members of the Euphorbiaceae, for example *Tragia ramosa* and *T. saxicola* (Thurston, 1976). Here, the syringe cells (which are subepidermal in origin) contain large crystals with sharp tips, elongated bodies, and basal fin-like protrusions. These missile-shaped crystals look ideal to be shot at high velocity into skin. One might imagine that they could be very painful, although the few *Tragia* stings I have encountered in the field do not feel very different from those of the common nettle.

Sting chemistry

Stings are where chemical and physical defences meet. Whereas some have been described well in terms of their structures they have not been studied in adequate detail at the molecular level. Not all reports concerning sting chemistry are of good quality, but among the few things that are clear is that many stinging plants inject an acidic solution into the skin. One can get the impression from the literature that formic acid is a major sting irritant.

However, one of the most credible reports of sting contents in the nettle *Urtica thunbergiana* from East Asia found otherwise. The stings of these nettles are small, containing an average volume of only 4 nL of stinging fluid (Fu et al., 2006). Nevertheless, enough of this fluid was collected and pooled, and found to contain on average 5 ng of oxalic acid and 57 ng of tartaric acid per sting. A further compound detected was histamine (21 ng per sting), a chemical which had already emerged from several older reports on the sting contents of nettles in the Urticaceae. Histamine, unlike tartaric acid and oxalic acid, is not part of primary metabolism in plants but it has numerous activities in animal cells including functions in neurotransmission and in mediating vasodilation and inflammation. After identifying and quantifying the major compounds in the stings of *U. thunbergiana* the work was then continued. Oxalic acid and tartaric acid were both able to produce persistent pain in assays using the hind paws of rats, but formic acid, a minor component of the sting fluid, was less active in eliciting the pain sensation. Histamine may accentuate the pain caused by the organic acids by stimulating swelling and by facilitating the dispersion of these acids into tissues. There may be other mediators from stings that have specific biological activity in humans. These include acetylcholine and even serotonin (5-hydroxytryptamine; Thurston and Lersten, 1969), although the evidence for the presence of these compounds in plant stings is still fragmentary. Finally, there are multiple mentions of the polypeptide content of stings in the literature (Thurston and Lersten, 1969) and perhaps proteins also influence the sting sensation.

What are stings directed against?

The four sting-producing plant families have one thing in common—as flowering plants go, they are relatively modern. Assuming that stings have been gained rather than lost during evolution, this relative modernity, coupled to the fact that stings might not be particularly effective against hardened animal body parts, would be consistent with them evolving as defences against herbivores that radiated relatively recently. This fits in well with most of the available evidence nearly all of which points to stings being deterrents to vertebrates. In the late 1880s Stahl observed that stinging hairs did not stop the snail *Helix pomatia* from feeding on the shoots of *Urtica dioica* and concluded that stings were defences against 'higher' animals (Stahl, 1888). Similarly, many insects lay their eggs on nettles (Wheeler, 2005), suggesting that stings might not be effective barriers to most invertebrates. There are, however, accounts concerning plants in families outside the Urticaceae giving a different perspective. Mala mujer (*Cnidoscolus urens*, Euphorbiaceae) is a highly defended spurge that is armed with stings and poisonous latex. These plants, growing in northern Costa Rica, were reported to trap insects on their leaf stings. To overcome these stings, the larvae of a specialist moth on the same plant bites them off at their bases, thereby gaining access to the leaf into which it then digs channels that will block latex flow to their feeding sites on the leaf (Dillon et al., 1983). Clearly, insects that are specialized to feed on mala mujer can deal with its stings.

Let us now return to Stahl's claim concerning higher animals. Roles as deterrents against vertebrates have been borne out by experiments using common nettles (*U. dioica*) as well as wood nettles (*Laportea canadensis*). None of the four types of invertebrate tested (a beetle, a butterfly larva, a grasshopper, and a snail) were deterred by the stings of the common nettle and this led the authors to conclude that they were more likely to have evolved as defences against back-boned herbivores (Tuberville et al., 1996). Previously, the finding of common nettle populations with particularly low densities of stings on their leaves (Pollard and Briggs, 1984a) had made it possible to test whether or not stings reduce feeding by sheep and rabbits. In both cases the herbivore preferred to eat the low sting variety over the more common higher sting density nettle (Pollard and Briggs, 1984b). Nevertheless, a few clever vertebrates can get around stings. Mountain gorillas in Rwanda, for example, eat nettles by rolling up the leaves so that the stings are crushed or occluded before they hurt the mouth (Tennie et al., 2008). But the bulk of observations to date suggest that strong selection pressures for sting evolution have come from large animals. I doubt if anyone would disagree with this if they had been stung by the ongaonga.

Unless one is very unlucky, one only touches the ferociously powerful and well-named ongaonga (*Urtica ferox*, Figure 3.4B) of New Zealand once in one's life. A single sting is so powerful that it would seem unlikely that it had evolved against anything but a hefty vertebrate, something bigger than a human. This is almost certainly the case, the ongaonga being one of the few stinging plants for which we can be almost certain about the target herbivores: the now extinct moas (Atkinson and Greenwood, 1989). Consistent with this, endemic stinging plants tend to be absent from the parts of many islands that were probably never reached by large vertebrate folivores, the Juan Fernández archipelago in the eastern Pacific being a good example. Curiously, however, there is a stinging nettle, *Laportia interrupta*, on New Caledonia, an island for which there was little impact of indigenous vertebrate folivores. One obvious origin for this non-endemic plant could have been a recent accidental introduction by Europeans. However, at least one specimen in the herbarium in Noumea is dated 1855, so it is possible that the plant was on the island prior to European arrival. With this in mind I asked the opinion of Bill Sykes, a doyen of Pacific Island botany, who told me that *L. interrupta* had been used as a food by Melanesians. This nettle could, then, have been trafficked on to New Caledonia by the island's early settlers.

Stings are under-researched structures and this is a pity. The Loasaceae of western South America with their beautiful flowers, plants from the Urticaceae like the tropical American cow-itch (*Urera baccifera*), the North Indian *Laportia crenulata*, and, of course, the New Zealand ongaonga and the Australian Gympie nettle have much to offer in the search for molecules or even proteins that might have uses in targeting pain-transducing pathways in humans.

Glochids, hairs, prickles, and thorns

The smallest manifestations of spininess are the hardened hair-like structures that can make leaf surfaces feel prickly. Some of the best known and most irritant examples come from cacti such as prickly pears (*Opuntia* spp.; Cactaceae) and these are termed glochids, a term that is typically used only in relation to cacti. Glochids are tiny, often barbed, and they are borne in tight clusters, frequently between thorn areoles. They are one part of a two-tier defence system, the thorns being repellents and the glochids becoming a useful defence when the flattened *Opuntia* pads are bitten into or tugged. In this case, the glochids begin to detach themselves and are easily embedded in the soft tissues (tongues or eyes) of vertebrate herbivores. But could these small structures also defend against invertebrates? Perhaps glochids reduce the ability of small organisms to colonize thorn clusters. Nevertheless, a more even distribution of glochids on the pad surface would seem better suited to keep invertebrates at bay. Moreover, in some parts of the New World prickly pears are a source of food for tortoises, and it seems quite likely that both the thorns and the glochids on these plants are directed at these reptiles. This would make sense; easily detachable irritant hairs are found elsewhere in the plant kingdom (for example in some members of the Malvaceae). Glochids are sometimes so easily detached that they might function to some extent as analogues to the defence hairs on some large spiders which are typically directed against predators with soft body parts such as noses or eyes.

The terminology

Moving now to larger defensive structures, the fact that there is relatively little morphological constraint on thorn or prickle production is reflected in the great diversity and wide distribution of these structures. But as with several other defence features of plants there are tricky issues to confront when it comes to terminology (Bell, 1991). Here, thorn and spine are used as synonyms for sharp structures on plants that are either entire organs e.g. modified branches of blackthorn, *Prunus spinosa* (Rosaceae), or parts of organs (e.g. modified stipules on many acacias) or entirely modified leaves (e.g. thorns in most cacti; Figure 3.5A). The word prickle is used here for hardened, pointed emergences upon organs where the organ itself is not otherwise modified and retains its primary function—for example, prickles on the stems of roses (*Rosa* spp., Rosaceae) or brambles (*Ribes* spp., Rosaceae).

Thorns and prickles can have multiple roles

Before interpreting the potential roles of thorns or prickles in plant defence, one can first ask what other functions these structures could have. In some plants, prickles have roles largely restricted to climbing over other vegetation. For example, many rattan palms, most of which are tropical climbing plants, have two types of armament each with its own role to play. As illustrated in Figure 3.5B, they produce hooked prickles on apical, whip-like climbing organs (cirri or flagellae), and, lower down on their more rigid stems, rattans usually have purely defensive straight thorns that are of no use whatsoever for climbing. Similarly, whereas functions in defence are likely for the thorns in many cacti, they might also aid in trapping layers of humid air over the plants, or in filtering out ultraviolet light. So there is no reason that individual thorns or prickles could not have multiple roles. However, many of these structures have clearly co-evolved with vertebrate herbivores. Like stings, thorns target faces and in particular mouths. The hooked thorns of some acacias and cacti probably target tongues and lips. Thorns and prickles are often coloured and, at least from a human perspective, they often appear to be made to be seen. Many are white, as was

Figure 3.5 Thorns and prickles can have multiple functions. (A) Like most plants in its family, the cactus *Eriosyce chilensis* from Punta Los Molles, Chile, has leaves that have been reduced to thorns. The primary role of these thorns is defensive, but other functions in photoprotection or humidity control are possible if not likely. Photo: S. Rasmann, University of California—Davis. (B) Thorns and climbing emergences on a rattan (*Calamus* sp.) from northern Queensland, Australia. The plant has lower stems that are heavily armoured with defensive thorns. A cirrus is seen in front of the hand in the photograph. These elongated aerial stems are covered in hooked prickles that are used to help the rattan climb among vegetation.

mentioned previously for most African acacias. Some, like those on the climbing plant *Capparis sinaica* (also known as *C. cartilagnea*, Capparaceae) on the Arabian Peninsula are vividly coloured—in this case, orange.

Plant society effects on spininess

Whether a plant is likely to be spiny often depends on the nutritional value of its leaves and, critically, on the relative nutritional value of neighbouring plants. The analogy is simple. If you wish to burgle a house you will not want to waste time in fruitless infractions where there is nothing worth stealing. Instead, you may look for signs of wealth. If all the nearby houses look the same except for one that stands out by being surrounded by barbed wire, then it might well have more-valuable-than-average contents. In plants, spininess can be an indication of relative nutritional wealth. Peter Grubb used the nitrogen content of leaves to help explain patterns in the occurrence of spininess. Several examples from forest ecosystems illustrate this. Plants growing in the shade of the forest often have tough, long-lived, and nutrient-poor leaves. But gaps, created when large trees fall, are niches for fast-growing plants with nutrient-rich, short-lived leaves. The leaves of these gap-requiring plants can contain almost twice as much nitrogen as the leaves of the shade-growers and the gap colonizers in some forests also have a tendency to spininess. Montane forests of Papua New Guinea display this relationship. Only 1.3% of shade-tolerant trees, shrubs and climbers in these forests were spiny but the number increased to 28% for similar plants growing in gaps (Grubb, 1992). Having leaves that are richer in nutrients than one's neighbours' leaves is risky, so stronger defensive measures than those used by the neighbour are required.

Spininess and the environment

Anyone who thinks that spiny plants are found only in arid or windy environments should try touching the underwater surfaces of the Amazonian water lily *Victoria amazonica* (Nymphaceae). Whereas small children can be placed comfortably upon the upper surface of these lily pads, this would definitely not work if the floating fronds were inverted. Prickly emergences coat all the major water-exposed surfaces of these giant water lilies and they are big enough and sharp enough to have evolved against sizeable swimming vertebrates, perhaps some of the larger rodent species that inhabit the river basins and flood plains of South America, perhaps herbivorous fish. In humid forests in Central and South America many large-trunked tree palms are spiny, as are some completely unrelated dicotyledonous climbers, e.g. some *Smilax* species (Smilacaceae). Finally, in Britain, the marsh thistle (*C. palustre*) requires damp habitats. But there is often an association of spininess and climate severity. The difference in this case is that large numbers of species from diverse families can be armed. This occurs in parts of the Sonoran desert extending from southwestern USA into Mexico, in some steppe systems in southeastern Europe, and in the dry spiny forests of Madagascar, etc. In a classic and attractive paper Rauh (1943) came to three conclusions about the use of thorns for plants. These were: (i) to protect the plant from excessive water loss through transpiration; (ii) to facilitate the capture of water vapour; (iii) to protect the plant from being eaten by animals. The order of these conclusions should be reversed.

The middle zone

A first conclusion is that finding individual spiny plants in otherwise non-spiny floras often relates to increased nutritional value relative to the neighbouring plants. But when large numbers of species in the same flora bear spines these plants often occur in dry or windy environments where there is always a strong history of browsing. Abiotic stresses can increase the propensity for spininess and this occurs in part because they act to restrain plant growth, keeping the plant body small and within reach of the herbivore. At the one end of the spectrum of plant communities are dense forests of large trees and where the bulk of green mass is inaccessible to vertebrates. At the other end of the spectrum would be climate extremes such as in alpine zones. There, a low plant mass per unit surface area coupled to a low herbivore density means that pressure from vertebrate herbivores may be reduced relative to more favourable climates. So, the leaves on large trees found at the base of a mountain may show relatively little sign of physical defence against herbivory. Similarly, there are few or no spines on cushion plants high in the alpine zone. But moving down the environmental gradient into the subalpine zone brings more warmth, more plant growth, and potentially, more herbivory.

A high level of defence expenditure can be predicted for intermediate zones, regions where plants grow to a relatively high mass, but where there may be environmental constraints on their height. When the whole plant body throughout each life phase is accessible to dominant herbivores then one can expect good defences. Extending this to ecological scales may help explain why there are large regions on earth where one finds common defence syndromes: zones of spininess, floras with chemical defences, or floras with a strong component of mimicry and crypsis. Such regions are typified by vast areas of thorn cushion extending from parts of the Mediterranean through to Anatolia and into Iran—a flora characterized by genera such as *Astragalus* and *Onobrychis* (both Fabaceae), *Acantholimon* (Plumbaginaceae) and *Acanthophyllum* (Caryophyllaceae). The aromatic Mediterranean or Californian floras that are often dominated by Asteraceae and Lamiaceae, the remarkable spiny flora of parts of the Mohave desert, the visual crypsis in the Namaqualand flora, the *Berberis* and *Colletia*-rich spiny undergrowth common to parts of the foothills of the Andes, etc. It is in the middle zone where, in addition to plant–plant competition, growth is restricted by a combination of biotic and abiotic factors, where one often finds plant communities that invest heavily in physical and chemical defences.

Defensive thorns have not evolved in the absence of vertebrate herbivores

Thorns evolved to inflict pain in vertebrates but it can be difficult to know which ones. The difficulty in understanding the defensive utility of thorns for individual plant species was captured by the Swiss scientist Jean Senebier in 1800: ‘One can’t believe that spines are the plant’s arms against animals because donkeys eat thistles and this is where goldfinches find their seeds, and caterpillars find refuge and food among spines’ (Senebier, 1800). Senebier never looked deeply into the question of spininess in plants, but if he had, he may have encountered difficulty attributing a defined defensive role for the prickles on a thistle leaf. It’s a bit like trying to ask why porcupines are spiny. Their spines could have evolved under pressure from a number of different predators. This might also be the case for thistles. Take, for example, those of the thistle *Cirsium spinosissimum* (Asteraceae), a mountain plant that grows in the Alps not far from where Senebier lived and worked. What exactly are these thistles defended against? It could be one or all of the following: chamois, ibex, snow hare, marmot, voles, or perhaps even large gastropods. The difficulty is that it may have been several of these animals or perhaps another animal that is now extinct. But clues exist around us: even if thistles are eaten by donkeys, they are often avoided by cattle. Fortunately there are better examples where we can strongly associate the evolution of thorns to particular herbivores or suites of herbivores.

The places that had few vertebrate herbivores until they were brought by humans usually lack indigenous plants that carry defensive thorns. A good example is the island of Juan Fernández, and another island with an even bigger flora is New Caledonia. From the faunistic history of both island systems where, at least in forests, terrestrial vertebrate folivores appear to have had little impact on the flora, one would not expect to find spiny plants. This is indeed the case and there are indeed remarkably few such plants on these islands. On New Caledonia the approximately 20 endemic *Pandanus* (Pandanaceae) species have prickly leaf-edges. But for all the dicotyledonous plants there are only two or three endemic species found in drier coastal regions that have backward-pointing thorns that would aid in climbing or sprawling (or possibly in defence against long-extinct tortoises). Indeed the plants, *Caesalpinia schlechteri* (Caesalpinaceae) is a sprawler and *Capparis* sp. (Capparaceae) a liana. Three other non-endemic but native plants possess thorns: *Ximenia americana* (Olacaceae) has occasionally been found as a spiny form, but even then its thorns are fairly small and widely spaced. *Machlura cochinchinensis* (Moraceae) is a climber/sprawler with straight thorns ~1 cm long and *Carissa ovata* (Apocynaceae) has pairs of thorns, also ~1 cm long, at its leaf bases. Thus, apart from the native *Pandanus* species, only these three indigenous but non-endemic plants have anything resembling defensive thorns, remarkably little spininess for a flora of more than 3000 vascular plant species.

Monkey puzzles

To further test the thorn–herbivore association one can approach this from the opposite direction and ask whether non-spiny plant genera on New Caledonia have spiny counterparts elsewhere? A good example is from conifer genus *Araucaria*, the genus to which the monkey puzzle (*A. araucana*) of South America belongs. Not one of the 13 species of *Araucaria* in New Caledonia, all of which are endemic, is spiny. The same is true for the attractive Norfolk Island pine, *A. heterophylla*. The paucity of spininess in New Caledonia and Norfolk Island araucarias would be consistent with the near absence of recent indigenous macroherbivores. By contrast, the five other *Araucaria* species in the world are all at least to some extent spiny. Australia has two species, *A. bidwillii*, the bunya pine and *A. cunninghamii* var. *cunninghamii*, the hoop pine. Bunya pine is spiny and hoop pine is somewhat so; certainly new shoots that emerge at ground level from the hoop pine can be extremely spiny. Then there are two species in Papua New Guinea, *A. cunninghamii* var. *papuana*, which, like its Australian counterpart, is moderately spiny, and *A. hunsteinii* that looks like a spinier bunya pine with slightly bigger leaves. The two remaining species occur in South America; the monkey puzzle itself, growing on the border of southern Chile and western Argentina, and *A. angustifolia* which grows in parts of Brazil, northern Argentina and Paraguay. Together, these are the two spiniest members of the genus.

This now raises the question of which herbivores selected the well-defended leaves on the five *Araucaria* species that still live outside New Caledonia and Norfolk Island. Unfortunately, this is difficult to answer. For a start, the spininess of the two South American species differs from those living in the Australo-Papuan region; the trees almost certainly evolved under different herbivore pressures. It would seem logical that the South American araucarias protected the plant from climbing or gliding intruders that would eat the leaves, break into the cones, or simply damage branches—as do birds and roosting fruit bats. However, based on the near lack of thorns in terrestrial mammal-free New Caledonia (where birds and fruit bats exist) the South American herbivores in question probably were, or are, mammals. In South America there are folivores such as monkeys, as well as opossums and also squirrels and other rodents, all of which could be potential attackers of the monkey puzzle. Australia and New Guinea both have marsupial herbivores including possums and tree kangaroos as well as a few rodents, but they have no native primates. Thus, in common between Australia, New Guinea, and South America there are marsupials and (to a different extent in Australia) rodents. Based on the fact that marsupials generally make better arboreal leaf eaters than rodents, the name 'marsupial puzzle' would seem at least as suitable as 'monkey puzzle'. It would pay off to pursue the monkey puzzle puzzle in the fossil record.

Suites of large herbivores can select for physical defences

Thorn size, structure, ontogeny, and placement pattern all reveal much about the type of herbivores that were likely to have selected for these defences. In many cases thorns may have evolved against a sufficiently broad range of herbivores so that their size and placement is likely to be the best achievable solution to these differing herbivore pressures. For example, most spiny African acacias are covered from top to bottom in similar-sized thorns, and it is sometimes said that the thorns of these acacias are defences against giraffes. This may be too simplistic: the acacias growing in East Africa may be attacked from the near ground level by dikdiks, at intermediate level by various medium-sized gazelles, at still higher levels still by antelopes that stand up on their hind legs and use their long necks to reach foliage high above the ground. Then, at the level of their crowns, giraffes may eat the trees. To add to this, vervet monkeys may supplement their diets with foliage high in the trees during the day to be replaced at night by galagos (bush babies) that also eat some leaves. So the majority of the vertebrate herbivores that feed on African acacia trees are ungulates. This group (as opposed to a single species) is likely to be the major driver of thorn selection in these trees, and in a beautiful example of co-evolution there is a lock-and-key fit between the muzzles of some gazelles and interthorn spaces in acacias. This is the case for at least two of the Horn of Africa's gazelline antelopes, the gerenuk and the dibatag.

There are many other examples where the co-evolution of plants with discrete groups of herbivores is likely to have led to spininess. In some cases thorns have clearly evolved under pressures from discrete groups or 'suites' of herbivores. Murids are among the major herbivores on palms and these rodents particularly like to eat young shoots and ripening fruits. The thorns and prickles on some palms are quite likely to be antirodent defences. Indeed, roots modified as thorns are frequently produced on trunks and in some palms they are oriented downwards and sufficiently closely spaced to block the passage of animals of mouse and rat size. Again, reinforcing the vertebrate–thorn relationship, whereas the majority of palms in western Malaysia produce prickles on their stems, none of the 32 palm species in New Caledonia do so. This is a place where terrestrial rodents and other ground-dwelling mammals have historically been absent from the fauna.

Germ warfare using thorns?

Thorns can be coated in a diverse microbial flora and it has been suggested that thorn-borne, disease-causing bacteria may augment a plant's defence. Several pathogenic bacteria including

Bacillus cereus and *Clostridium perfringens*, causal agents of food poisoning and gas gangrene respectively, were indeed found on the sharp tips of young date palm (*Phoenix dactylifera*, Aracaceace) leaves (Halpern et al., 2007). However, other parts of the plant surface near the leaf thorns were not used as controls in this study, and they too are probably colonized by these rather common bacteria. It would not, then, be unexpected that some thorn-caused injuries later led to disease, but any infection would likely develop too slowly to be associated with learning. In spiny plants the deterrent is the physical pain sensation. The herbivore does not even have to break into a leaf to encounter this. But if it does, it will face other physical defences most of which are of cellular or subcellular scale.

Silica: targeting teeth and mandibles

Herbivores need to keep their teeth or mandibles in good condition in order to fully extract and digest cell contents efficiently. Furthermore, they need to avoid crushing plant cells that border feeding wounds since this can activate defence gene expression in plants. So it would make sense if plants had ways to wear down mandibles or teeth. An ideal abrasive would have to fulfil a number of criteria. It would be easily available, simple to incorporate into or on to tissues, and at least as hard as enamelled mammalian teeth or chitinous insect mandibles. Furthermore, it would have to be resistant to degradation by enzymes secreted by the herbivore. Finally, in the plant, it would not block valuable light from penetrating the leaf. This is asking a lot for one substance—but glass fulfils all these demands. Glass is essentially what is used in the plant kingdom: hydrated silicon oxide otherwise known as silica (Epstein, 1999). The acquisition of silicic acid and its incorporation as silica is energetically inexpensive and offers an excellent possibility to plants to augment their defences. But, before looking at this, one can ask whether there are plants that forgo the accumulation of soluble silicic acid and use sand grains directly.

Sand armour

The capture of sand and soil has been discussed in terms of crypsis (Wiens, 1978) but it would make sense that some plants might also have evolved ways to catch sand grains and to coat their surfaces with this cheapest and hardest of defences. If so, a good place to begin to look for such putative defences would be where sand is abundant—the Arabian Peninsula, for example. There are indeed good examples of sand coatings in this region. Grasses like *Stipagrostis* often have exposed roots coated in sand and plants in several dicotyledon families (Nyctaginaceae, Caryophyllaceae and, above all, the Molluginaceae) on the Peninsula catch sand grains on their stems. But could sand grains be used as a physical defence for leaves? Surprisingly, this type of defence strategy has not, to my knowledge, been discussed as such. If this occurred one would expect that too much sand on the upper surface of a leaf might reduce light capture since sand is almost invariably coloured. Therefore, fully coated upper leaf surfaces might be rare. By contrast, sand grains could theoretically be trapped on the undersides and edges of leaves provided that they do not interfere with gas exchange. To look into this we undertook a hunt for sand-catching plants in Saudi Arabia.

Figure 3.6 An example of sand armour. Sand armour on the desert plant *Indigofera argentea* (Fabaceae), Jizan, Saudi Arabia. Stems, leaf edges, and the undersides of leaves are covered in regular-spaced sand grains. Scale bar = 1 cm. (See Plate 7.)

It did not take long to find such a plant on sun-baked apricot-coloured sand just north of the steamingly hot port city of Jizan on Saudi Arabia's southern-most Red Sea coast. *Indigofera argentea* (Fabaceae), with its grey-green foliage and bright red flowers, was growing as widely spaced and easily seen clumps about 15–25 cm in diameter. Most of the pinnate leaves of these small plants (Figure 3.6) were found to have five to seven leaflets about 3 mm wide and 4 mm long and each covered in white, hair-like trichomes with irregular surfaces. The trichomes on the stems and leaflets trapped sand grains of various sizes. Counting only the sand grains with diameters that exceeded those of the trichomes from ten leaflets from this plant gave the following distribution: an average of seven grains on the upper (adaxial) leaflet surface and 31 grains on the lower (abaxial) leaflet surface. Additionally, grains were found around the leaf edges and, interestingly, they tended to be evenly spaced along these borders. One does not have to go to Saudi Arabia to find such sand coatings and a plant growing on coastal sand in Wales at Ynyslas appears to do what *I. argentea* does. Colonies of common rest-harrow (*Ononis repens*) growing on bare dunes (that are colonized by rabbits) had sand coatings on their upper stems and, above all, on the lower surfaces of the young leaves. Underscoring the fact that defence strategies (or in this case putative defence strategies) tend to occur repeatedly in the same family, this plant is, like *Indigofera*, in the Fabaceae.

Silica capture and deposition

To build tailored silaceous defence structures on or within the body, plants first take up silicon from the soil in the form of silicic acid $Si(OH)_4$ which is then loaded into the xylem from which it is transported to shoots to be deposited as insoluble silica (Ma and Yamaji, 2006). Silica deposition is highly utilized in some early plant groups such as horsetails (*Equisetum* spp., Equisetaceae) as well as in several modern angiosperm families, most notably the sedges (Cyperaceae) and grasses (Poaceae) and in several tropical families where it can be so abundant in wood that it can quickly blunt saws (e.g. trees in the Chrysobalanaceae). Examples of large families that are intermediate for silicon acquisition capacity are the nettles (Urticaceae), the spiderwort family (Commelinaceae) and the cucumber family (Cucurbitaceae). To work optimally as a defence, silica deposits must be positioned correctly and macroscopic examples of this are easy to see. For example, the saw-like, silicon-rich teeth on leaf edges in razor sedges and razor grasses are arranged so that they can best protect the central growth regions of the plants. Human passers-by can receive 'paper cuts' from these plants but these leaf edges are far worse for organisms that try to reach into the centre of tussocks. Razor-edged leaves may have evolved to damage the soft muzzles and mouths of mammalian herbivores that try to reach young leaves. But it is smaller, silica-rich structures within leaves that abrade teeth.

Phytoliths and geladas

Phytoliths, crystal-like deposits of silica, are found widely in the leaves of many plants, and perhaps most notably in grasses. Like grains of sand, they resist decomposition and can survive in soil for millennia, and their distinctive shapes allow paleobotanists to identify the species from which they originated (Piperno, 2006). In the living plant these structures occur in vacuoles or cell walls and sometimes silica depositions are so concentrated that whole cells become silicified to become 'silica cells.' Phytoliths and silica cells can be beautiful objects, although one usually needs some patience to isolate them from fresh plant material. However, an easy way to enrich for phytoliths, fibres, and other indigestible matter is to examine the faeces of grass eaters. This can be done with geladas, the world's only primate grazer, now only found in some upland parts of Ethiopia. When dry, the surface of gelada faeces becomes water repellent and almost shiny; this is consistent with the fact that they are impregnated with silicon-rich structures that pass from grass leaves unharmed through the entire digestive system. In the laboratory it is easy to bleach and destroy the matrix of microbial matter and phenolic compounds, leaving a delicate white suspension dominated by fibre cells from about 150 to 500 μm in length and about 10 μm in width. Beneath the microscope other structures become apparent. Xylem vessels with well-preserved spiral secondary thickenings are visible as are phytoliths and silica cells, both of which are shown in Figure 3.7.

Silica deposition in grasses has been found to be induced by herbivory (McNaughton and Tarrants, 1983) and silica levels in the tufted hair-grass (*Deschampsia caespitosa*) in northern England have been found to correlate negatively with the density of field voles (*Microtis agrestis*), a rodent that uses *D. caespitosa* as a winter food (Massey et al., 2007, 2008; Massey and Hartley, 2009). Through tooth wear, silica in leaves can reduce herbivore lifespans, but some animals counter this. For example, in Australia the narbalek, or little rock wallaby (*Petrogale concinna* Macropodidae), eats an aquatic fern (*Marsilea crenata*, Marsileaceae)

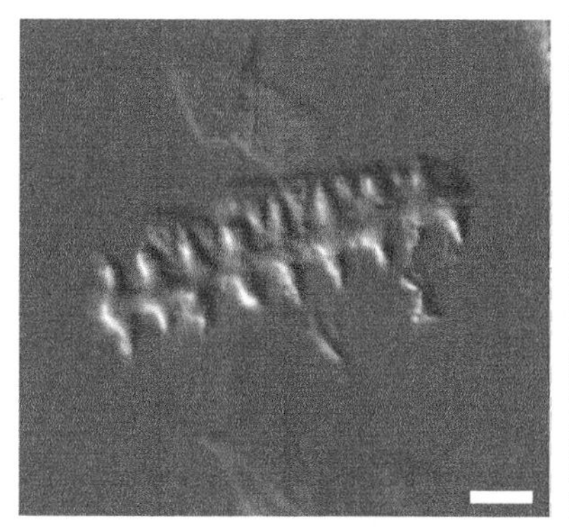
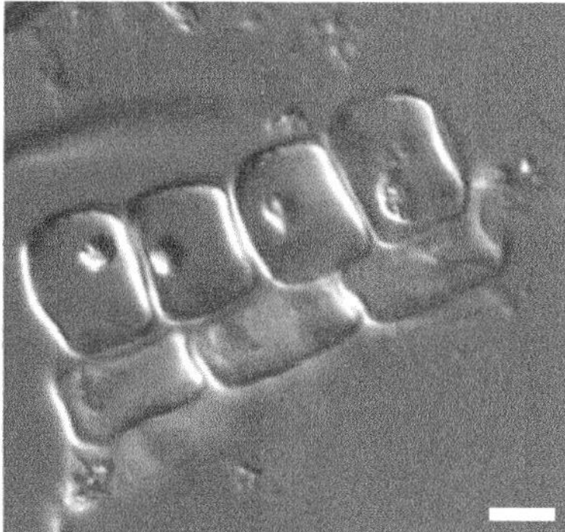

Figure 3.7 Silica-rich defence structures from grasses. Phytoliths and silica-based cell wall reinforcements resist digestion in the guts of specialist grazers like the gelada (*Theropithecus gelada*). To isolate the structures, dry faeces (25 mg, from samples collected in November 2010 in Debre Libanos, Ethiopia) were extracted by incubating in a mixture of concentrated HNO_3 (3 mL) and 50% H_2O_2 (2 mL) for 5 days at room temperature. The precipitate was washed in water and suspended in 70% (v/v) glycerol for observation by differential interference contrast microscopy. Scale bars = 20 μm.

containing >20% dry weight silicon oxide. It is the only macropod to have constantly growing teeth. Turning to invertebrates, some leaf-cutting ants that harvest silicon-rich leaves even reinforce their mandibles with zinc (Schofield et al., 2002) and the giant New Caledonian grasshopper (*Pseudophyllanax imperialis*), a truly formidable insect with a wing span of up to 22 cm, can slice through the tough, fibrous and silica-rich leaves of banana and coconut palm.

Crystalline defences

In addition to the use of silica in protection against herbivores, the leaves of many plants contain chemically simple but structurally diverse crystals that serve similar purposes. Botanists have given these crystals a bewildering variety of names including prisms, styloids, druses, raphides (needle-like crystals often packed into clusters), and crystal sand (Figure 3.8). The bulk of these solids contain calcium, usually as calcium oxalate (Franceschi and Horner, 1980; Prychid and Rudall, 1999) and sometimes as calcium carbonate (Arnott and Pautard, 1970; Nitta et al., 2006). Crystalline deposits of calcium carbonate occur, for example, in several members of the Cannabinaceae (e.g. hop, *Humulus lupulus* and *Cannabis sativa*). These plants have epidermal 'lithocysts', a specific term for cells that contain crystals of calcium carbonate (to make matters more complicated these calcium carbonate crystals are themselves termed 'cystoliths'). The leaves of figs (*Ficus* spp., Moraceae) are also well known for their epidermal lithocysts. But, on the whole, calcium oxalate crystals are more common in leaves than are those made of calcium carbonate.

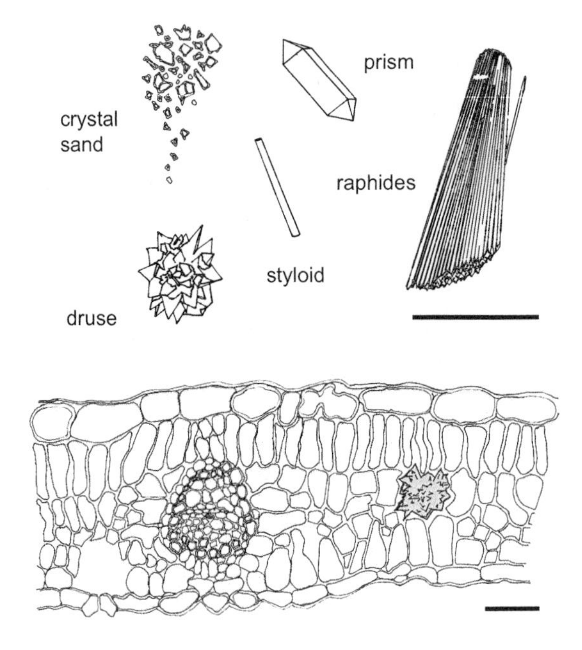

Figure 3.8 Crystal deposits common in leaves. (*Upper*) Various crystal types. Scale bar = 50 μm. (*Lower*) An extracellular calcium oxalate druse (grey) in the leaf of the pedunculate oak (*Quercus robur*). Transversal section, scale bar = 300 μm.

Crystals of calcium oxalate can be found within vacuoles and sometimes in the cell walls or extracellular spaces in leaves, as in the pedunculate oak (*Quercus robur*), which has large extracellular druses in the upper spongy mesophyll layer in leaves in late summer (Figure 3.8). Some crystals are even intracellular and extracellular at the same time, for example the prismatic crystals in the leaves of the water hyacinth (*Eichhornia crassipes*, Pontederiaceae). These look like tiny javelins that have been fired through leaf parenchyma cells. As in the case of other small defensive structures the placement of crystals in the leaf is optimized. In many plants the calcium oxalate crystals appear to protect the major photosynthetic cells, either the palisade layer or the mesophyll or, in other cases, they are placed proximally to the vasculature. All these patterns occur in the Piperaceae, a family in which several genera produce spectacular arrangements of crystals (Horner et al., 2012). These are easily overlooked under standard light microscopy but they appear brightly in polarized light (Figure 3.9).

The association of leaf patterning and strong defences

There seems to be an association of leaf surface colour patterning and the accumulation of silica or crystal deposits in or on leaves. The leaves of numerous plants that contain crystal idioblasts, particularly those from tropical regions, display striking visual patterning. This is

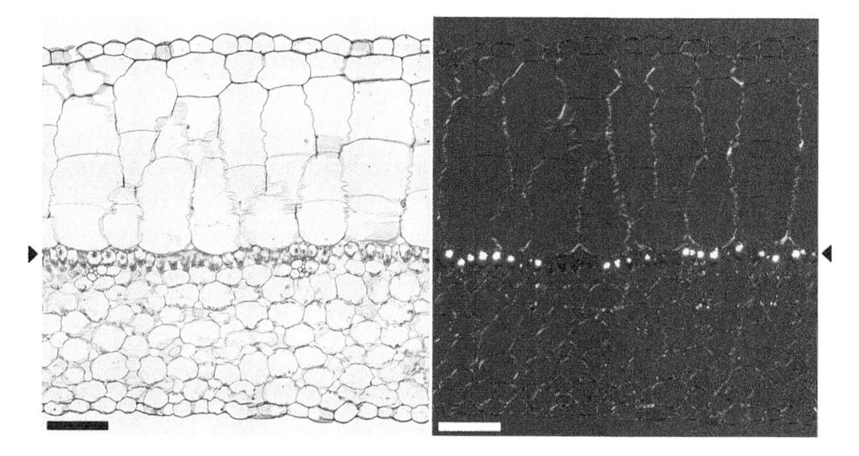

Figure 3.9 Intracellular calcium oxalate crystals in photosynthetic cells in a leaf of *Peperomia prostrata*. The image on the left was viewed in white light. On the right is the same image viewed under polarizing light. The arrowheads show a discrete layer of crystal idioblasts containing druses. Scale bar = 150 μm. The majority of leaves contain at least one type of strategically placed idioblast. Image: G. Colau.

the case in genera such as *Peperomia* (Piperaceae), *Begonia* (Begoniaceae), and *Dieffenbachia* (Araceae). Crystal production reaches its zenith in this latter family, the aroids, for example in the spectacular genus *Caladium* with its crystal-rich and remarkably decorated leaves. An example in temperate regions is the common lungwort (Figure 2.6) with its stiff, silica-rich trichomes. The association of leaf patterning with physical or chemical defences would be consistent with pattern use for warning. Patterning and crystal production is not restricted to the tropics. The barrel medic (*Medicago truncatula*, Fabaceae), a now widespread plant originally from the Mediterranean region with striking crimson markings on the surfaces of each leaflet, is another calcium oxalate accumulator. The sheaths of cells surrounding secondary veins in the leaves of this plant contain large prismatic calcium oxalate crystals. Lines of *M. trunculata* lacking these crystals were more readily eaten by beet armyworms (*Spodoptera exigua*) than were those of the wild type (Korth et al., 2006). Furthermore, the mandibles of armyworms that had been reared on the wild type were more worn down than those of the insects reared on calcium oxalate-containing plants. More recently, the ability to make calcium oxalate crystals resembling druses has been engineered into *Arabidopsis* (Nakata, 2012); this opens up the possibility of penetrant studies of crystal defences and it should soon be feasible to determine their effects on mammalian herbivores.

The fact that crystals are easily observed makes them attractive for a wide variety of experiments. One may ask: if there are populations of crystals in different cell types, which population is lost first when access to a component of the crystal is reduced through mineral depletion? This has been done with the leaves of the Chinese mulberry (*Morus australis*, Moraceae) which have both epidermal cells containing calcium carbonate crystals and bundle-sheath cells containing calcium oxalate. Since both types of cell contain calcium deposits their content of this metal is subject to its availability in the soil, this raised the question of which cell population sacrifices its crystals when calcium becomes limiting? To test this, small *M. australis* plants were starved for calcium and compared with plants grown under non-limiting levels of this metal. In the calcium-starved plants the epidermal lithocysts developed normally but contained reduced levels of calcium carbonate, whereas the bundle-sheath crystal idioblasts completely lacked calcium oxalate crystals (Wu et al., 2006). Defences are probably organized in hierarchies with some being prioritized over others.

Fibre cells and sclereids

A great deal of the defensive capacity of leaves is chemical in nature. However, as plants start to mature and approach the reproductive phase, the levels of defence chemicals in their leaves often decline. These leaves will not necessarily become any easier to eat, since protective physical structures may replace the chemicals, so as one type of defence subsides another replaces it, leaving few windows of opportunity for the leaf eater. In the case of defence mechanisms in plants, chemical defences have traditionally stolen the limelight from their physical counterparts. From this perspective the contributions of two types of cells—fibre cells, and sclereids—deserve attention. At the outset, and before discussing fibre cells, the word 'fibre' needs defining since 'dietary fibres' and 'fibre cells' are not the same thing. Dietary fibres are polymers (i.e. chemical entities) that resist digestion in humans and they are classified as being either insoluble or soluble. The main examples of the former are cellulose and hemicellulose; an example of a soluble fibre is pectin. These are often only poorly degraded by herbivores without the additional help of enzymes produced by their symbionts. Unlike cellulose that can at least be partially degraded by many specialized herbivores, lignin, a polymer based on phenylpropanoid units, has evolved to be almost indestructible. It is an inaccessible source of carbon for all herbivores. Fibre cells and sclereids are both heavily lignified and can contribute heavily to the leaf defence arsenal, impeding the digestion of otherwise available food matter (Van Soest, 1996).

Fibre cells

Fibre cells (Evert, 2006), otherwise known as sclerenchyma, are elongated and mechanically strengthened due to heavy lignocellulosic thickening of their walls. Furthermore, individual fibre cells are often found packed into bundles, making them very resilient superstructures that add greatly to the amount of lignin contained in leaves. These cells are often associated with the xylem but, in some plants, extraxylary fibre cells can give tissues even more fibrous characteristics. Some of these individual fibre cells can be up to several tens of centimetres long, as in agaves (*Agave* spp., Agarvaceae), the New Zealand hemp (*Phormium tenax*, Agarvaceae), and sansevieras (*Sanseviera* spp., Asparagaceae). The long fibre cells in these plants probably grow by what is known as intrusive growth, an uncommon growth process in plants involving expansion of the whole cell surface, not just the cell tips, so that the elongating cell essentially invades groups of other cells (Chernova and Gorshkova, 2007). Another common arrangement of fibre cells is as sheaths around individual vascular bundles; these represent formidable barriers to sucking insects such as aphids. Unlike xylem, fibre cells do not necessarily die once their development has been completed. Most remain alive and in a seemingly quiescent state, sometimes containing starch granules (Murphy and Alvin, 1997; Evert, 2006). This is an excellent place to store and protect starch: structural cells can protect food sources.

Fibre cells are not typically considered as defences; their primary role is clearly mechanical and their placement in many leaves attests to this. For example, grass leaves tend to have many bundles of fibre cells in the long axis of the leaf and placed on the abaxial surface (underside) near the epidermis. These cells help support leaves, but in addition they contribute to make the leaves hard to chew. The types of leaves typically rich in sclerenchyma cells include many that we humans would not think of eating: palms, grasses, sedges, agaves, etc., and those of many conifers. It is therefore not surprising that some plant families containing members that often invest heavily in chemical defence may include one or a few species that have little chemical defence but which, instead, are extremely fibrous. Ramie (*Boehmeria nivea*), a plant in the Urticaceae, does this. Another spectacular example of the use of fibre cells is found in a large shrub growing in sand from parts of the Arabian Peninsula to India: *Leptadenia pyrotechnica* (Apocynaceae). Whereas most plants in the same family growing in these regions are good latex producers but are not particularly fibrous, *L. pyrotechnica* produces no visible latex but has an extremely high fibre cell content (explaining why it has been used for millennia to start fires). This plant is almost tasteless but is next to impossible to chew and swallow.

Sclereids

A second type of cell that shows many parallels with fibre cells is the sclereid: a compact non-elongated cell with thick extracellular matrices and sometimes bearing sharp projections. These cells are widely distributed and extremely variable in shape and size (Evert, 2006). Like fibre cells, they almost certainly play as much of a role in defence as they do in general tissue strengthening. At the borderline between fibre cells and sclereids would be filiform sclereids such as those found in the mesophyll cells of olive (*Olea europaea*, Oleaceae). At the other extreme, sclereids may be almost spherical such as stone cells ('brachysclereids') from the flesh of pear (*Pyrus communis*, Rosaceae) fruits. Almost every possible intermediate form of these cells seems to have been made somewhere in the plant kingdom, from highly branched stellate structures to angular cubic shapes or the spiked, club-like sclereids found at the end of vascular bundles in the leaves of some Australian plants such as *Hakea* (Proteaceae). Another almost stellate sclereid type is from the fragrant water lily *Nymphaea odorata* (Nymphaeaceae) from North America. In this example there are crystals embedded in the outer walls of these structures. At present there is little evidence for roles of these cells in any aspect of plant physiology (for example, in water transport), and it is most likely that all these cells help to defend plants from herbivores.

Defence cells: idioblasts

The term 'idioblast' refers to individual cells that are, quite simply, different from their neighbours (Evert, 2006). For example, crystal-containing cells are 'crystal idioblasts'. There are many other types of idioblast and they can be found throughout the plant, being in the internal layers or in the epidermis. Many have strange shapes and interesting contents, and they are positioned strategically within tissues where they function by concentrating defence substances, or enzymes that are necessary for the release of active defence chemicals from conjugates, usually glucosides. Some idioblasts have structures that permit the targeted delivery of their contents to specific parts of attacking organisms. For example, *Dieffenbachia* leaves contain multiple types of crystal with the most remarkable being raphides which are stored in idioblasts termed biforines. These raphides are expelled rapidly if the cell is damaged (Coté, 2009).

Finding novel idioblasts, even in highly studied plant species, is not uncommon. The cotyledons and mature leaves of *Arabidopsis thaliana* contain idioblasts called myrosin cells and these, like stomatal guard cells in *Arabidopsis*, store the protein myrosinase. Myrosin cells have been known for a long time, and are not difficult to find, being associated with vascular strands in the leaf (Ueda et al., 2006). However, many idioblasts are visually cryptic; they may closely resemble surrounding cells under the microscope, but differ markedly in cell contents. This is the case for glucosinolate-containing idioblasts that were found by scanning transversal sections of *Arabidopsis* flower stems for their elemental composition. Within what looked like a field of parenchyma were cells that were highly enriched in sulphur: S-cells. When longitudinal sections of the stem were made, these revealed the S-cells to be extremely elongated and quite unlike their neighbours—they are glucosinolate-containing idioblasts (Koroleva et al., 2000). Being in some cases rich stores of nutrients, it is necessary to ask whether idioblasts have roles in plant growth. There is, however, little indication of this. For example, experiments aimed at genetically ablating specific idioblast populations have been performed with oilseed rape (*Brassica napus*, Brassicaceae). When these plants were engineered so that they failed to produce a particular idioblast population in their cotyledons the plants still grew normally (Borgen et al., 2010). There is at present far more evidence for the roles of idioblasts in defence than in growth.

Exudates

The term idioblast is usually restricted to single cells. More complex multicellular defensive structures retain liquid chemical defences known as exudates, fluids that drip from plants or accumulate on the surfaces of leaves and other organs. The main categories of exudates are gum, mucilage, resin, and latex. Phloem sap is not included in this list. Unless channelled out of plants by sucking insects, phloem sap does not drip out of plants. Indeed, phloem cells have rapid mechanisms for sealing themselves if they are damaged (Van Bel, 2003). Only in some exceptional cases, such as in certain genera in the Cucurbitaceae, do extrafascicular phloem-like cells produce exudates that are almost certainly defence-related. Exudates are chemical mixtures and this brings us a step closer to looking at defence at the molecular level. But before this, more emphasis will be placed on chemical storage compartments that are often made to rupture upon attack.

Gum and mucilage

Gum and mucilage are both based on water-soluble polysaccharides. In the case of gum the polysaccharides originate from cell wall breakdown making them extracellular. So gums are found in cavities or pockets between living cells. A good example of this would be gums produced by many *Vachellia* (*Acacia*) species (Fabaceae). These seep out of wounded trunks in sometimes prodigious quantities, trickling downwards to harden and to look deceptively like resins. Kino is another type of gum, a mixture of polymerized phenolics and polysaccharides produced by many *Eucalyptus* (Myrtaceae) species. Closely related to gums are mucilages which differ from gums in that their principal components are not derived from cell wall breakdown and they are usually stored within specialized cells such as idioblasts or trichomes.

Resin ducts and laticifers

Resins are lipid-soluble mixtures based on phenolics and/or terpenes, pine resins being classic examples (Langenheim, 2003). These fluids are often very rich in volatile organic components such as monoterpenes and they are low in water-soluble compounds and proteins. This, from a biological point of view, is remarkable. Resins are, essentially, bio-incompatible, and many are highly inflammable. Not surprisingly, these fluids have to be produced by specialized cells from which they are secreted into the lumen of ducts, canals (elongated ducts) or cysts for storage until the plant is damaged. One of the most spectacular flowering plant families for resin production is the Burseraceae, the family in which both frankincense (*Boswellia*) and myrrh (*Commiphora*) trees are found. The plants that produce these types of exudate can be very strongly defended indeed. In the case of large numbers of species within this family (trees in the Burseraceae can be dominant species in some tropical regions) damage causes resins to trickle out of the trunks of these trees rather slowly. But several *Bursera* species squirt their resins for several tens of centimetres. They can literally blast insects off leaves (Becerra et al., 2009).

Whereas resins are essentially organic solvents, latexes are water-rich emulsions. Although the main components of latexes, which are highly polymerized terpenes, can be lipid soluble, the water content of latex ensures that they remain in suspension rather than in solution. Furthermore, being in an aqueous environment facilitates the production and storage of proteins in latex. So latex can be seen as being more biocompatible than is resin and, not surprisingly, latexes are stored inside living cells: laticifers (Pickard, 2008; Agrawal and Konno, 2009). Laticifer cells can be remarkably long, sometimes measuring many centimetres, and, as with any other defence-related cells, they are positioned strategically in the plant to protect sites of vulnerability. Both laticifers and resin ducts often follow the venation, and feeding herbivores are likely to break them, causing the rapid release of their contents. To circumvent this, specialist insects that feed on latex- and resin-producing plants have evolved clever strategies to disarm the defence. Many of these insects will actively cut through laticifers or resin ducts prior to feeding on leaves. The way they do this is quite remarkable because of the danger involved.

Laticifer architectures range from simple branching patterns following the venation pattern of the leaves to complex, net-like arrangements (articulated anastomosing laticifers). Insects that are specialized to feed on branched latificers cut through one or more of these so that latex is released upstream of where they will feed. By contrast, those that eat leaves with net-like laticifers have to cut trenches to isolate a region in which they can feed safely (Dussourd and Denno, 1991). Similar phenomena occur in resin-producing plants and the invertebrate folivores that feed on them have to be very specialized. The more defence cells that the plant uses, the more specialized an insect has to be. But even if a herbivore is specialized to sever laticifers or resin ducts, this process costs time, thus increasing the danger of being discovered by predators. The sophisticated adaptations of specialized invertebrate herbivores beg the question of whether vertebrates might have adapted to dealing with exudates. Remarkably, some of them actually feed on these fluids.

Exudivores: defence substances as food

Mammalian exudivores are among the most fascinating of herbivores and they come from groups as disparate as possums in Australia, flying squirrels in the Americas, some lemurs in Madagascar, and galagos in Africa. The diets of these mammals can be strange indeed—in some cases they consist mostly of gums. For example, in Australia the yellow-bellied glider (*Petaurus australis*) cuts incisions into bark to feed on *Eucalyptus* kino. A second possum, the sugar glider (*P. breviceps*), itself too small to penetrate the coarse bark of many trees, often acts as a secondary feeder, exploiting the damage left by yellow-bellied gliders. These and many other exudivores exploit a food source that presumably first evolved as a defence against invertebrates and pathogens.

Interestingly, a high proportion of exudivores also include leaves in their diet. Towards the leaf-eating extreme of the entire spectrum of possums are members of the ringtail possum family (Petaduridae), most of which are highly folivorous. The green ringtail possum (*Pseudochirops archeri*), where it has been studied in detail, shows a particularly strong preference for the leaves of four seemingly diverse and unrelated tree species, these making up more than half of its diet. These trees are *Aleurites rockinghamensis* (Euphorbiaceae), *Ficus fraseri* and *F. copiosa* (Moraceae), and *Arytera divaricata* (Sapindaceae) (Jones et al., 2006). What they have in common is that they all produce latex. As Andrew Krockenberger and colleagues have pointed out (Jones et al., 2006), the green ringtail possum is likely to be a dietary specialist that reduces the diversity of plant defence chemicals it ingests by selecting latex-producing trees. Exudivores illustrate the fact that some animals feed on substances that probably evolved in the first place as plant protectants. Animals that specialize highly on plants sometimes become reliant on their defences.

Summary

Leaf defence against herbivores rests to a large extent on specialized cells (trichomes and idioblasts) as well as on macroscopic structures such as stings, prickles, and thorns that have evolved to inflict pain in soft-bodied organisms. Many individual defence cells and cell clusters are easily visible to the naked eye, as are trichomes, structures with multiple established roles in defence. But often hidden from sight are mouth- and mandible-damaging abrasives such as silica or sharp crystals. Most leaves employ defence cells for the storage of chemicals (or of enzymes that activate defence chemicals) and having specialized storage cells and storage compartments offers many advantages, allowing plants to greatly extend the range of chemicals that they use in defence. Even when only found in small quantities in leaves, defence chemicals are rarely spread evenly throughout plant cells or through all cell types (e.g. Shroff et al., 2008). Instead, they are often concentrated in specialized cells that are positioned to optimize leaf defence. For protection, the strategical organization of defence cells in the inner cell layers of the leaf is just as critical as leaf surface topology.

4

Chemical defences

Being autotrophic, plants have the raw materials, the energy, and the reducing power to perform very complex chemistry. Many species exploit their synthetic chemistry capacity to the full and they often couple this to the use of specialized chemical storage cells. Viewed as a whole, the plant kingdom produces tens of thousands of different chemical structures and each plant species contains a small spectrum of these. But this chemical blend differs between organs and it is also influenced by plant age, environmental factors, and disease or predation. How has the chemical diversity found in the plant kingdom arisen? In addressing this question, we can ask whether there are any universal defence chemicals that are made and used in all land plants?

A few chemicals do indeed appear to be produced very widely, particularly in wounded plant tissues. These include green leaf volatiles (GLVs) such as 3*Z*-hexenal and its isomer 2*E*-hexenal, molecules that are generated from oxygenated fatty acid precursors. The almost universal occurrence of hexenals throughout the plant kingdom was discovered long ago, and the authors of a notable paper suggested that they might be a kind of diffusible disinfectant produced to protect plant tissues from pests and pathogens (Schildknecht and Rauch, 1961). Today, we know that there is some truth in this, but GLVs have a wide variety of roles not all of which are related to defence against biotic stresses. However, it is hard to name other examples of anything approaching a universal low-molecular-mass defence chemical. Plant chemistry, then, is staggeringly complex (Wink, 2008). Importantly for leaf survival the ability of land plants to fabricate and sequester chemicals of various sorts is greater than that of the animals that attack them.

Selection for novel defence chemistries

In one of the most highly cited papers in the field of co-evolution, Ehrlich and Raven (1964) explained how plant chemicals act in nature as trophic (feeding) barriers to insects, in this case lepidopterans. The idea is that overcoming trophic barriers due to a plant's particular chemistry has led to herbivore radiations. This view is based on the fact that each plant species has its own cocktail of defence compounds, meaning that there are almost as many blends of chemicals in plants as there are plant species. This makes it difficult for herbivores that pass much or all of their time on one host to switch to a different plant species where they would need a different battery of enzymes to detoxify a different spectrum of defence compounds.

The only way that herbivores can evolve to cross chemical barriers is by adapting to the new plant in a reciprocal evolutionary process which should lead to radiations of herbivores and hosts. Consistent with this, insect clades containing herbivores are particularly diverse compared with sister clades of non-herbivorous insects (Mitter et al., 1988; Wheat et al., 2007). In fact, Ehrlich and Raven's paper provided an explanation of why there are so many herbivorous insects and so many plant species. It has since spawned a large literature related to the important question of biological diversity and, while there is still debate about how important plant chemicals really are in promoting insect speciation, evidence that is consistent with this keeps on accumulating (Mauricio and Rauscher, 1997; Fine et al., 2004; Agrawal, 2005; Futuyma and Agrawal, 2009). A good way of visualizing the process is to think of each plant as an island.

The concept of trophic islands: examples from New Caledonia

Daniel Janzen and others have considered plants as islands whereby colonization by herbivores requires adaptation to the specific conditions on each plant species (Janzen, 1968), just as it might be necessary for humans to adapt to life on a particular island. In the evolutionary course of specialization to a particular plant diet the herbivore must gain the capacity to detoxify, sequester or void toxic chemicals. Furthermore, the analogy can be used because the sort of chemistry used by plants tends to be conserved within families so that most genera and species produce chemicals that fall into particular conserved classes. In this view, if individual species are considered as trophic islands then plant genera would be archipelagos. In large part due to leaf chemistry, it is easier for a herbivore to hop from island to island in an archipelago than to move between archipelagos. I had never fully appreciated this until visiting New Caledonia and witnessing the effect of trophic barriers on both humans and plants.

New Caledonia is situated in the South Pacific >1000 km from the east coast of Australia. It is a large island (19,060 km^2) that has a tremendously rich and unusual flora with about 3000 native species of vascular plants. Of these, just over 2300 are listed as being endemic to the island and a high proportion of this endemism is at the level of genera and families (Jaffré et al., 2001). The island also has a relatively rich extant reptile fauna, but none are known to be obligate folivores. There are birds such as the notou, a giant pigeon that inflicts a modest amount of damage to leaves, but the only native mammals are fruit bats which have had, at most, a limited impact on foliage. As on any other island the faunal history of New Caledonia has changed over time and some of its animals have died out. Extinct land tortoises, for example, may have had some impact on the flora. Also notable was at least one species of the large flightless bird *Sylviornis* (Balouet and Olson, 1989), although there is at present little evidence to suggest that this bird was a folivore. Things changed precipitously with the arrival of humans at least 3500 years ago when our species began the process of introducing herbivorous mammals, a process that was accelerated dramatically with the immigration of Europeans less than 300 years ago. Fortunately, Europeans documented the distribution of the tribes of earlier Melanesian settlers and noticed a striking pattern that is still seen today.

Whereas certain parts of the main island were and still are well populated, other large areas were never settled. This unsettled surface area is itself large (> 5000 km^2), amounting to one-third of the island's total. What typifies the unpopulated areas is that the soil is ultramafic; it is a trophic island within an island. The ultramafic soil is a barrier to human food production and, related to this, it has allowed the evolution of plants that are themselves trophic barriers to insects. Ultramafic soils are typically poor in minerals essential to plant growth, such as phosphorus, potassium, and calcium. By contrast, these soils are rich in iron, magnesium, nickel, chromium, cobalt, and manganese. Evidently, it is difficult to grow domesticated plants on this soil, yet the ultramafic regions are well vegetated and have floras rich in endemic species. This is a remarkable place to see evolution in action and to find interesting adaptations in plants and herbivores.

As the entomologist Jean Chazeau (1993) has written:

> The ultramaphic marks a frontier. Humans, who are pioneers particularly tolerant of difficult environmental conditions, have never established themselves there. . . . To prosper in this environment requires overcoming a trophic barrier. For a herbivore, overcoming this barrier is necessary for the exploitation of the plant host. To be successful requires two evolutionary processes; acquirement of characters of tolerance to a poor nutrition and, secondarily, the acquisition of resistance characteristics to a potentially toxic resource. As for plants, the trophic barrier is an efficient filter to the installation of competing species and it thus facilitates the success of pioneer species and radiation of these species, radiation amongst lepidopterans is thus concentrated on the ultramaphic soil and not on the sedimentary soil.

What makes this relevant from a defence point of view is that, in many cases, the chemistry typical of a particular plant genus has changed if it has speciated on the ultramafic soil. This is the case, for example, in *Psychotria* (Rubiaceae) where several species growing outside the ultramafic region contain high levels of alkaloids (as is common in the Rubiaceae). But the endemic *P. douarrei* which grows on ultramafic soil contains up to 5% dry weight nickel and has low levels of alkaloids. In this case nickel, a toxic transition metal, appears to have replaced nitrogen-containing chemical defences (Davis et al., 2001). In other words, plants growing on heavy-metal-rich soils can both economize on using nitrogen in defence chemicals and can exploit a new chemical diversity by mining the soil.

At least 65 plant species from a variety of plant families on New Caledonia are Ni hyperaccumulators and the island is ideal for their study (Jaffré et al., 2013). But surely the most remarkable of these plants is the endemic nickel tree, *Sebertia acuminata* (Sapotaceae). At first sight there is nothing startling about this tree and one would walk past it without thinking unless one knew what was beneath its bark. However, Tanguy Jaffré discovered that *S. acuminata* has a latex containing a staggering 26% dry weight of nickel (Jaffré et al., 1976). The nickel tree is adept at concentrating this metal. It grows on soils typically containing only 0.1–0.3% dry weight nickel, so this represents a roughly 200-fold concentration of the metal

Figure 4.1 The nickel tree, *Sebertia acuminata*. Cuts into the bark cause the release of a blue latex that contains 26% dry weight nickel. The laticifers containing this latex extend into the leaves. (See Plate 8.)

compared with concentrations in the ground. When a cut is made in the bark, the cream-blue latex that is enriched in nickel citrate (Sagner et al., 1998) oozes out (Figure 4.1). The leaves of this tree contain about 1.2% dry weight of nickel, and this is a substantial amount since most of it is concentrated in laticifers. Additionally, a small amount of the metal is detectable on the leaf surfaces. Even bryophyte epiphylls on *Sebertia acuminata* and *Psychotria douarrei* are nickel-rich.

How might invertebrate herbivores deal with potential nickel poisoning? Some clues to this are found by inspecting the Ni-accumulating fruits of *P. douarrei*, only one of several endemic *Psychotria* species on the island that accumulate this metal (Jaffré et al., 2013). I frequently observed the larvae of small moths within the berries of *P. douarrei* and when the insects were squeezed gently they appeared to excrete the easily detected metal, that is they probably have the ability to void the nickel rather than taking it up. Without this adaptation the berries would almost certainly represent a trophic barrier to the moth. What, then, does nickel citrate do for the nickel tree *Sebertia acuminata*? Inspection of the bark of some of the larger nickel trees reveals numerous superficial galleries presumably made by insects. But it is harder to find any of these galleries that penetrate deeper into the tree. It may be that the tree's latex protects against bark-boring beetles, folivores, and perhaps numerous other invaders, including pathogens.

Other elemental defences

Metal-accumulating plants occur fairly widely and most species that do this are specialized to hyperaccumulate one of a large number of elements including copper (Cu), zinc (Zn) and cadmium (Cd), and, of course, nickel (Ni), etc. As more examples are discovered there is growing evidence that this phenomenon plays roles in defence against pathogens and herbivores (McNair 2003; Rascio and Navari-Izzo, 2011). In addition to the deployment of metal ions some non-metals are also mined and accumulated to high levels in leaves where they act as antiherbivore defences. One example is selenium (Se) accumulation in several milkvetches (*Astragalus* spp.; Fabaceae). Although selenium is not stored in plants in its elemental form (it is stored as selenocysteine, a modified form of the amino acid cysteine), it sheds further light on leaf defence. *Astragalus bisculcatus* frequently grows around black-tailed prairie dog (*Cynomys ludovicianus*) colonies in the western USA. On Se-rich soils this plant is reported to accumulate remarkably high levels of Se, up to 10 grams per kilogram fresh weight. Prairie dogs avoid eating high Se individuals of this plant (Quinn et al., 2008).

There is a periodic table of elements used in plant defence chemistry, and many elements are exploited. Before discussing the various groups of plant chemicals built from these atoms it is worthwhile to consider why some elements have or have not been used. For example, the use of nitrogen in defence merits attention since this element is valuable and sometimes growth-limiting. However, plants have the ability to embed nitrogen into larger molecules in ways that make it particularly hard for a leaf eater to extract, and this has facilitated the evolution of a large variety of exquisitely specific molecular interactions between plant chemicals and diverse target molecules in herbivores. But whereas nitrogen-containing defence molecules are common in the plant kingdom, those that contain phosphorus are very rare. The probable explanation for this is that phosphorus-containing molecules might give this valuable atom away because, in a biological context, phosphorus usually finds itself in phosphate groups that are esterified on to other molecules. These ester bonds, however, are readily broken down by hydrolysis and plants do not seem to have been able to exploit certain atomic linkages such as those in powerful synthetic pesticides such as organophosphates. Sulphur, like nitrogen, is relatively common in plant defence chemicals and it appears in many molecular contexts including in elemental form (Cooper et al., 1996). Many sulphur-based chemicals in plants are volatile and are released from leaves upon attack. A walk through a forest in Europe in the spring reminds us of this. The smell of wild bear garlic (*Allium ursinum*; Liliaceae) is largely the smell of breakdown products released from its pre-made sulphur-containing defence compounds (Schmitt, et al., 2005) as the plant is damaged by herbivores. In fact, it is the smell of plant defence in action.

Phytoanticipins

'For unfailingly secure defence, defend where there is no attack.' The words of Master Sun in Sun Tzu's *The Art of War* (Tzu, 1988 edition) can be used to emphasize the fact that much of a plant's chemical defence capacity is in place even in the absence of attack. This is the case for the majority of the compounds to be discussed in the following sections. The chemicals that are present in the unharmed plant are often termed phytoanticipins—they are there in anticipation of attack, or at least to form a barrier to non-specialized attackers. Although this term was originally used for antimicrobial compounds (VanEtten et al., 1994) it is increasingly used for antiherbivore compounds. However, the levels of many phytoanticipins, perhaps the majority, increase upon attack. For example, nicotine, an alkaloid produced by tobacco, is present constitutively in the aerial parts of the plant. But if insects attack tobacco leaves, more nicotine is synthesized in roots and recruited to the site of attack (Steppuhn et al., 2004).

Do chemicals really play a role in defence against herbivores?

It has been argued that evolving the ability to make large quantities of chemicals has been easy for plants because there are few constraints on this process (Carmona et al., 2011). But for the leaf, the elaboration of chemical defences can be costly, often using valuable nutrients either for building the chemicals themselves or in the construction of specialized defence cells. We can be certain that making defence chemicals often puts a burden on the plant and this is supported by studies with dioecious species, plants in which male and female flowers are borne on separate individuals. The different sexes have different tasks, with males having to produce lots of pollen and then females subsequently having to invest heavily in producing seeds. Producing copious amounts of pollen consumes nitrogen reserves (pollen grains are packed with nucleic acid and protein). In keeping with this, if males are compared with females at the pre-fertilization stage when the male is occupied with pollen production, the males in some plants are found to be more herbivore-damaged than are the females (Ågren et al., 1999).

Clearly, not all plant chemicals are defence-related. High-abundance chemicals in leaves could be storage pools of valuable elements such as nitrogen or they might have roles in the attraction of certain beneficial insects. Alternatively, they might provide protection against ultraviolet light or perhaps they have other roles as signals or toxins within or outside of the plant. There are examples of all of these in nature. However, it is likely that the majority of 'secondary' chemicals (i.e. those not needed for essential primary metabolism) are defence-related (Fraenkel, 1959). One of the best indications of this is the fact that specialized chemical storage cells made to rupture on attack have evolved repeatedly in plants. Nevertheless, alternative roles for all compounds mentioned herein should be sought: if a molecule has an antiherbivore role it could also be antimicrobial.

For the biologist, plant-derived chemicals are more interesting for what they do than what they are. Ideally, one would like to know which, if any, herbivores they evolved against, and how they function at the molecular and ecological levels. However, there are simply too many compounds, too many plants, and too many herbivores to allow this to be done systematically. Instead, much of what we know has come from experiments with invertebrate herbivores fed on plants in which the levels of chemicals have been altered genetically. In addition, the search for pharmaceuticals has revealed the physiological effects of large numbers of plant-produced chemicals on mammals including humans. But much less is known about which herbivores are targeted in nature by a particular defence chemical. This is in part because the ability to test the roles of chemicals in a biological context through genetically based experiments is new and work has been limited to only a few plant species. What has emerged, however, is that many if not most defence chemicals are not merely foul tasting or repellent. They appear to have evolved to interfere selectively with discrete physiological and behavioural processes in herbivores. Herbivores have to do their best to stop plant chemicals getting to their targets, and to do this they (and we) employ batteries of sensory receptors many of which have evolved to protect the animal from poisoning.

Chilis, capsaicin, and the initial detection of plant chemicals by herbivores

Herbivores and humans are well equipped to detect dangerous chemicals and to avoid them. This can be illustrated with a well-known first example from fruits rather than from leaves. Found naturally in dry regions such as Mexico and the Southern USA, chili plants (*Capsicum* spp., Solanaceae) have tiny seeds that are potentially vulnerable to destruction in mammalian digestive systems where food transit times are typically long. It is therefore important for the plant that the seeds are dispersed by organisms with digestive transit times that are faster than those found in most mammals. But desert mice can easily climb vegetation and could steal and destroy the chili's fruits. Capsaicin, the principal 'hot' component of the fruit, stops this happening (Nabhan, 1997; Tewksbury and Nabhan, 2001).

The levels of capsaicin are higher in the seeds than in the fruit flesh and the molecule is ideally placed for its role in nature—to protect the delicate seeds against rodent granivores. Capsaicin is so repulsive to mice that they will become severely malnourished rather than attempt to eat chili fruit or artificial diets containing capsaicin. Instead, the chili's seeds are dispersed by northern cardinals (*Cardinalis cardinalis*) and pyrrhuloxias (*C. sinuatus*), birds with rapid transit rates that do not destroy the seeds. But how can these birds tolerate the chili fruits that are so distasteful to mice?

Capsaicin works well as a mouse repellent since it is a potent agonist (activator) for the transient receptor potential V1 (TRPV1) calcium- and sodium-permeable channel which is part of the neuronal thermosensory machinery (Caterina et al., 1997). The activation of TRPV1 by capsaicin causes cation influx into neurons in mice and humans and this initiates a pain response similar to exposure to heat. Most mammals such as mice have a form of TRPV1 that is closely related in its amino acid sequence to that in humans and all these channels are capsaicin-activated. In contrast, and although the bird form of TRPV1 probably functions perfectly well in thermoregulation, its TRPV1 has an amino acid sequence different from that in mice and humans, and it does not bind capsaicin. When a cardinal eats a chili fruit it cannot experience the rapid burning sensation that we do.

TRPV1, the capsaicin receptor, is one of a large family of ion channels and receptors that are specific to animals. A major role of many of these proteins is to detect chemicals of plant origin. That is, TRPs and other proteins such as bitter receptors (TAS2R proteins) allow animals to detect and to avoid countless potentially toxic molecules or molecules that might need considerable energy to detoxify (Figure 4.2). For example, insects and humans detect electrophilic and reactive plant chemicals (such as isothiocyanates) using TRPA1

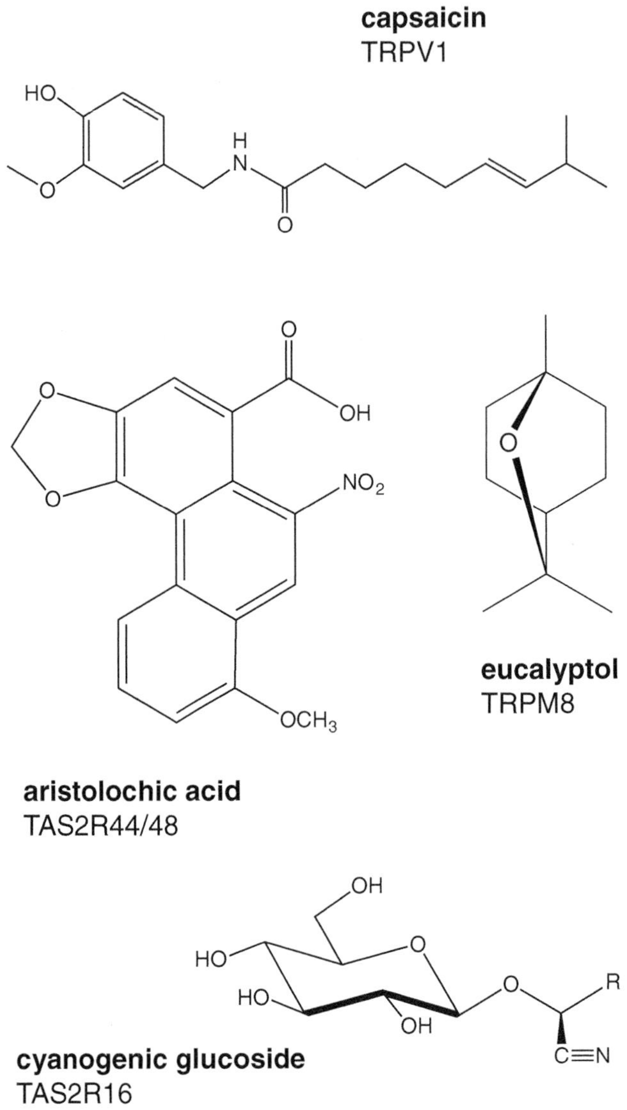

capsaicin
TRPV1

aristolochic acid
TAS2R44/48

eucalyptol
TRPM8

cyanogenic glucoside
TAS2R16

Figure 4.2 Examples of plant chemicals and their sensory receptors in humans. Many plant chemicals undergo at least three interactions with proteins in herbivores. They are first detected by sensory receptors in the mouth and digestive tract, e.g. by bitter receptors (TAS2Rs) or by transient receptor potential (TRP) proteins. In each case a principal sensory receptor from humans is indicated. Their detection in this way can lead to deterrence. However, if the compounds are ingested, many have other more specific molecular targets e.g. physiological receptors that function in the nervous system. These are referred to as the 'front-end' effects of the chemical. Finally, toxins from plants are detoxified in what are referred to herein as 'back-end' effects on the herbivore. All the compounds shown are from leaves except capsaicin which is from fruits and seeds of *Capsicum* (Solanaceae). Aristolochic acid is from plants in the Aristolochiaceae. Eucalyptol is a typical *Eucalyptus* (Myrtaceae) leaf monoterpene. Cyanogenic glucosides are made in a wide variety of families.

(Kang et al., 2010). A further related protein, TRPM8, is used to detect terpenes such as eucalyptol, menthol and menthone, all of which are widespread plant chemicals (Appendino et al., 2008; Vriens et al., 2008). The TRPs are one of our own strongest molecular connections to the plant world, many functioning in the digestive tract from the mouth to the anus (Holzer, 2011), constantly surveying the quality of what we swallow and try to digest. It is noteworthy that this entire class of receptor is lacking in plants, as are the genes that encode taste receptors such as TAS2Rs. Animals, then, use sensory receptors to help them avoid poisoning, but if chemicals evade this detection system they can reach other targets. This will now be illustrated, starting with nitrogen-containing defences and one of the simplest molecules among them: cyanide.

Cyanide

Roughly 3000 species of plants make cyanide precursors: cyanogenic glycosides (Zagrobelny et al., 2008). They occur in ferns such as bracken (*Pteridium aquilinum*, Dennstaedtiaceae) and in some gymnosperms including some *Taxus* species. In angiosperms they are common in many dicotyledon families (e.g. Asteraceae, Caprifoliaceae, Euphorbiaceae, Fabaceae, Linaceae, Myrtaceae, Rosaceae, Saxifragaceae, etc.) and in some monocotyledons but their production in a particular genus can be sporadic due to polymorphism, as for some of the cyanide-producing *Eucalyptus* (Mytaceae) species (Zagrobelny et al., 2008). Once released from its storage from, the cyanide anion, like many defence chemicals, has a discrete target. Cyanide targets complex IV of the mitochondrial respiratory chain, binding to cytochrome *c* oxidase and blocking the generation of adenosine triphosphate (ATP). Its generation from cyanogenic glycosides has been studied extensively in the seedlings of many of the cereals.

Dhurrin, the principal cyanogenic glucoside in sorghum, serves as classic example of how cyanide is stored and deployed by plants (Thayer and Conn, 1981; Vetter, 2000). It accumulates to high levels (up to 60 mg per gram dry weight) in the vacuoles of epidermal cells in the young leaves of seedlings. This is the phase during which the coleoptile emerges and the vulnerable first leaf then breaks through this protective sheath to begin expansion. At this point, cereal seedlings are especially vulnerable to many pathogens as well as to birds and rodents. Upon cell damage, hydrogen cyanide is released from dhurrin in an enzyme-catalysed process (Figure 4.3) where glucose is first hydrolysed off the glucoside by dhurrinases (β-glucosidases) to produce a hydroxynitrile. Then, in a second step, this is broken down by hydroxynitrile lyase to liberate hydrogen cyanide (HCN). Most dhurrinase activity is associated with chloroplasts in mesophyll cells whereas the hydroxynitrile lyase, also found in these cells, is mostly in the soluble fraction, probably in the cytosol. This is a typical arrangement for plant defence glycosides; the substrate is stored in a different subcellular localization (and often in a different cell type) to the enzymes needed to remove the stabilizing sugar. Given its mode of action on respiration—a deeply conserved molecular process—cyanide is a broad-spectrum toxin capable of poisoning microbial pathogens and herbivores as big as cows.

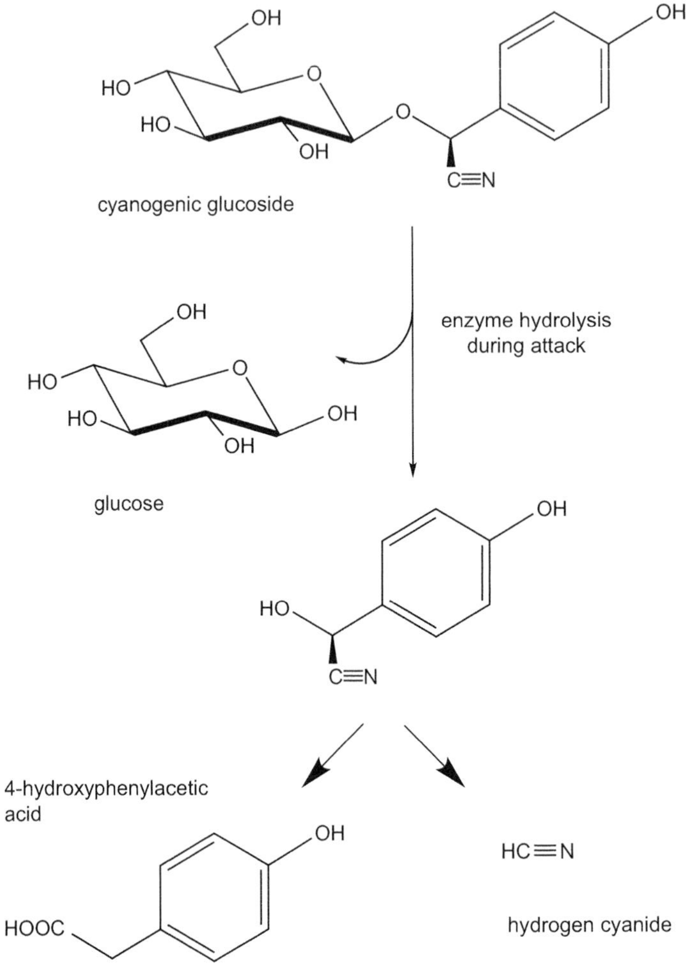

Figure 4.3 Production of cyanide from a cyanogenic glucoside.
Cyanogenic glucosides are hydrolysed upon attack by enzymes produced by the plant. This releases the glucose residue and generates an unstable intermediate that decays to release hydrogen cyanide, a cytochrome *c* oxidase inhibitor. The example shown here is dhurrin from sorghum.

Isothiocyanate

Another class of nitrogen-containing chemical in plants shows strong parallels to cyanogenic glycosides. These are glucosinolates; nitrogen- and sulphur-containing compounds that give mustard much of its distinctive flavour. Found in the plant order Brassicales and most notably in the family Brassicaceae, glucosinolates are a large and complex group of molecules that are collectively termed 'the mustard oil bomb' because they can be 'exploded' into a variety of small molecules when plants containing them are attacked. The mode of breakdown of

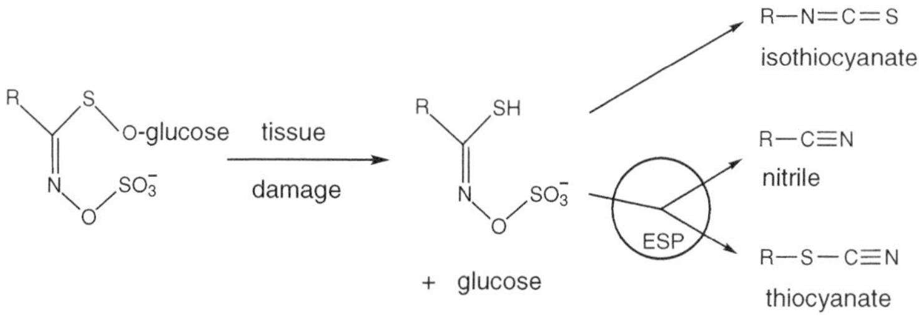

Figure 4.4 Glucosinolates: defence compounds typical of the Brassicaceae.
Glucosinolates, like many plant defence chemicals, are glucosides that are activated upon attack by glucose removal (catalysed, in this case, by myrosinases). The aglycone intermediate then decomposes to isothiocyanates, nitriles, and thiocyanates, and the proportion of these chemicals produced is controlled in part by plant-produced epithiospecifier (ESP) proteins. Isothiocyanates (ITCs) are likely to be the principal active compounds; they are electrophilic and readily bind to proteins. R is the variable part of the molecule.

glucosinolates upon attack is shown in Figure 4.4 and, as in the case of cyanogenic glucosides, it begins with the hydrolytic removal of glucose by plant-produced β-glucosidase enzymes called myrosinases (Rask et al., 2000; Wittstock and Burow, 2010).

Three main groups of glucosinolates exist: aliphatic glucosinolates, aromatic glucosinolates, and indole glucosinolates, and all of these can be broken down to liberate small molecules such as nitriles, thiocyanates, and isothiocyanates (ITCs) (Wittstock and Burow, 2010). The types and proportions of each of these products differ depending on the exact nature of the parent glucosinolate and can thus vary between individual species and even between different genotypes within the same species of plant. Fortunately, the apparent complexity of glucosinolate breakdown can be simplified: glucosinolates can be thought of as ITC precursors.

It is the ITCs that do most of the damage in the mustard oil bomb. In the reactive part of an ITC (Figure 4.4) is a carbon atom with two double bonds in the configuration (R-N=C=S where R- is a variable side chain). The central carbon atom in this context is electrophilic; it can react with amino groups or sulphur atoms in proteins and it is these sorts of reactions that presumably take place in the mouths and guts of herbivores to modify the activity of digestive or sensory proteins such as the receptor TRPA1.

Glucosinolates serve as a test ground for the antiherbivore roles of defence compounds. In *Arabidopsis* they are of two main types: alphatic and indole glucosinolates (Rask et al., 2000) and based on the fact that each of these two chemical subfamilies can be eliminated genetically, it is possible to test how each affects larval weight gain or insect reproductive efficiency. The growth of generalist insect herbivores is often strongly reduced by glucosinolates—but not always glucosinolates from the same class. For example, aliphatic glucosinolates in *Arabidopsis* reduce the growth of the tobacco hornworm (*Manduca sexta*), whereas green peach aphid (*Myzus persicae*) feeding on this plant is reduced by the indole glucosinolate branch. Both branches, however, reduce weight gain in the beet armyworms (*Spodoptera exigua*; Müller et al., 2010). In the case of the small cabbage butterfly (*Pieris rapae*), a specialist on Brassicaceae, this insect gains weight quite well on wild-type plants because it has clever ways of disarming ITC production. One of these mechanisms involves an insect-produced enzyme (glucosinolate sulphatase) that extracts sulphate from glucosinolates so that they simply cannot degrade to sulphur-containing molecules (Wittstock et al., 2004). Another mechanism uses insect-produced nitrile specifier proteins (NSPs) that, within the insect gut, redirect the breakdown pathway of glucosinolates away from ITCs and towards less toxic nitriles (Wittstock et al., 2004).

Several interesting challenges now remain. First, what are the principal molecular targets of ITC? This is still an open question although the answer is most likely to be proteins in the upper digestive tract. Another question is how do glucosinolates affect the feeding behaviour of vertebrates? This is still poorly known although the glucosinolates in healthy *Arabidopsis* leaves were not found to affect tortoise feeding (Mafli et al., 2012).

Alkaloids

Alkaloids are an ancient group of small, nitrogen-containing molecules most of which are derived from amino acids. They are found in some lycophytes, many ferns and conifers, and also angiosperms where they are more abundant in dicotyledons than in monocotyledons (major exceptions being Amaryllidaceae and Colchicaceae). Many of these molecules are potent toxins in humans but an even larger number might poison us or behave as drugs if our mouths and stomachs were not acidic. This is because most alkaloids have at least one amino group and these can be protonated and deprotonated depending on the pH of the solvent. In acidic stomachs these alkaloids are protonated and charged and this can reduce their uptake. Uncharged alkaloids can cross membranes more easily. For example, in humans cocaine is inefficiently taken up from the leaves of coca (*Erythroxylum coca*, Erythroxylaceae) unless alkaline substances are first chewed to reduce the acidity of the upper digestive tract (Davis, 1997). But the midgut of lepidopteran larvae is basic; many alkaloids can poison these insects efficiently.

Enzyme inhibition by molecular mimicry

Whereas many alkaloids target molecular signalling machinery (receptors, ion channels, protein kinases, etc.) in herbivores, others appear to have more prosaic roles acting as molecular mimics for central metabolites and inhibiting enzymes or metabolite transporters. For instance, the leaves of bluebells (*Hyacinthoides non-scripta*, Hyacinthaceae) in Europe contain alkaloids that inhibit various glycosidases and this is thought to underlie the digestive problems caused in cattle that eat these plants. Similarly, solanine from the leaves and greened tuber skins of potato (*Solanum tuberosum*) contain steroidal alkaloid glycosides called calystegines that can powerfully inhibit β-glucosidases (Asano et al., 2000). Swainsonine (Figure 4.5) from various members of the Fabaceae, including swainsona

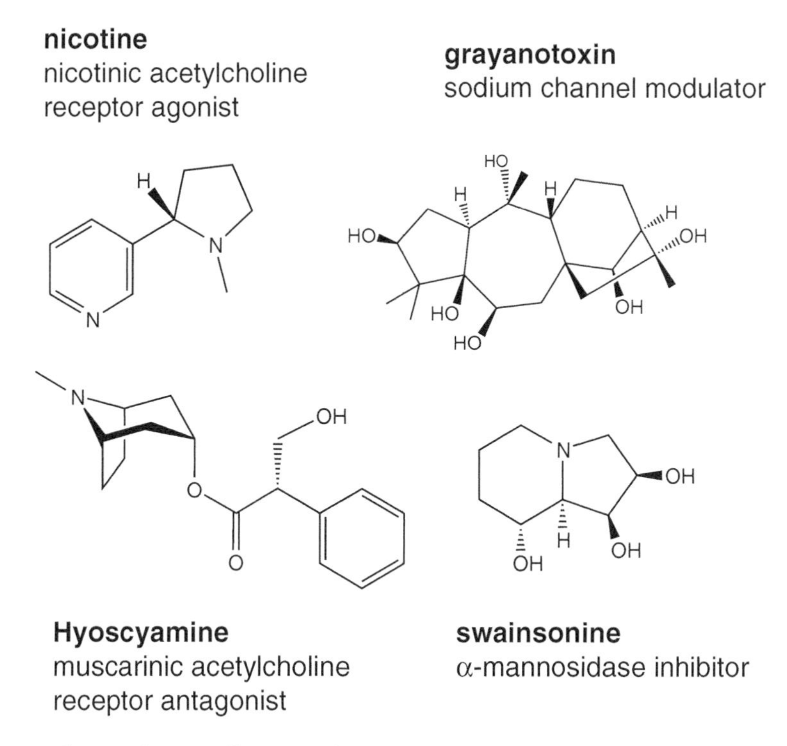

Figure 4.5 Leaf toxins that can affect animal nervous systems.
The compounds shown are alkaloids with the exception of a grayanotoxin, a diterpene that is a positive modulator of vertebrate sodium channel activity. The enteric nervous system (ENS) is a common target of plant defence chemicals.

(*Swainsona canescens*) in Australia and several *Astragalus* species in North America, is a toxic mannosidase inhibitor that through largely unknown mechanisms causes abnormal behavioural symptoms termed 'locoism' in livestock.

The dangerous effects of alkaloid overdosing

Interfering with the mind is a good way to put herbivores at risk and a chance observation made in northwestern Costa Rica bears this out. In the 1970s, K.E. Glander was patiently observing mantled howler monkeys (*Alouatta palliata*) in a leguminous tree (*Andira inermis*) when he saw a monkey, with her infant clutched tightly to her, begin to lose control of herself. Badly disoriented, she began to loosen her grip on the tree, eventually being held only by her tail until she eventually fell 10 m to the ground. The monkey was in some respects lucky. She was uninjured in the fall and did not suffer from predators. She later recovered and, still with her infant attached, made her way back up the tree where she remained more or less inert for a full day. Glander concluded that the monkey had probably consumed the alkaloid andirine in the *Andira* leaves (Glander, 1977).

This remarkable sighting led to Glander and his spouse spending more than 200 days studying the howler monkeys and their food trees. The monkeys were found to choose their food trees carefully and this was particularly evident for the madera negra tree (*Gliricidia sepium*, Fabaceae) the leaves of which contain several powerfully toxic molecules including rotenone and several alkaloids. The howlers fed preferentially on only three individuals although there were well over one hundred of these trees in the study area. Significantly, only the three that were fed on lacked detectable levels of alkaloids. The howlers often fed on the petioles of these trees, a type of behaviour seen in quite a few primates. Petioles typically have lower contents of defence chemicals than do laminas.

It is unfortunate that we do not know how the howler monkey felt before she fell, and our own perception of alkaloids is skewed by pleasurable effects of recreational drugs. But it is quite likely that the monkey felt very ill; in nature a primary target of many alkaloids is the digestive system and not the brain. Rather than creating euphoria, many alkaloids probably cause pain and digestive failure in herbivores by acting on the enteric nervous system (ENS). Either way the defence is good: if the herbivore is drugged it is in danger of injuring itself or being eaten by carnivores, otherwise its gut is poisoned.

Alkaloids from the Papaveraceae

Numerous plant-derived alkaloids are active on the nervous systems of both invertebrates and vertebrates with perhaps the most infamous coming from latex in the opium poppy *Papaver somniferum* (Papaveraceae). This latex, most concentrated in the laticifers in the periphery of the seed capsule (and also found in leaves and stems) contains five major alkaloids from two different structural classes; noscapine and papaverine from the one class; morphine, codeine, and thebaine from the other. Opiates have particularly powerful effects on nervous systems where they bind to opioid receptors (Waldhoer et al., 2004). Morphine itself has powerful effects on the gastrointestinal tract and can cause nausea, vomiting and constipation. But what is the role of poppy alkaloids in nature? Poppy latex is, in the first instance, a good deterrent to generalist insects that cannot disarm laticifers and that have not evolved to tolerate or detoxify the alkaloids. But it is possible that the effects of these alkaloids can spread even further. It is said that when locusts descend on opium fields to feed they become disabled and are fed on by rats which themselves suffer some of the effects of the alkaloids. The drugged rats may then fall prey to foxes. Apart from the few poppy species that produce morphine there are many other plants in the Papaveraceae that contain other alkaloids and these often seem to be well defended against vertebrates. One such plant is the especially invasive Mexican prickly poppy *Argemone mexicana*. Not only does it have thistle-like leaves, but it also contains toxic alkaloids. The fact that this plant thrives in warm climates along paths that are heavily used by cattle, donkeys, and camels shows that it has particularly robust antivertebrate defence. This is also the case for many plants in the Solanaceae.

Alkaloids from the Solanaceae

If a fast-growing plant is going to make high levels of alkaloids it helps to do what many toxic plants of the potato family do. They associate themselves with good sources of nitrogen (vertebrate herbivores) and then sequester nitrogen from the urine and dung of these animals to make potent defences that are targeted against these same animals. Deadly nightshade (*Atropa belladonna*) and henbane (*Hyoscyamus niger*) produce alklaloids that target nerve function. These chemicals can strongly affect the digestive tract—which may be their principal target. The digestive tract is surrounded by muscle and is highly innervated. Numerous alkaloids from the Solanaceae interfere with muscarinic acetylcholine receptor functions and can thereby affect both digestive tract muscle function and secretion. Scopolamine and hyoscyamine (Figure 4.5) from deadly nightshade and henbane are muscarinic acetylcholine receptor antagonists (Wess et al., 2007). One of their principal effects is to block secretion in the gastrointestinal tract. Vertebrates that consume these alkaloids can lose the ability to salivate and digest food properly or they become unable to move properly. If large quantities of these alkaloids are consumed, a herbivore can suffer deadly increases in heart rate. Nicotine (Figure 4.5), by contrast, is a nicotinic acetylcholine receptor agonist. Through targeting both nerve-muscle and nerve-nerve synapses it can block muscle function in the digestive tract and elsewhere in the body (Xiu et al., 2009). This is not its only effect and it has also been found to cause appetite suppression in mice (Mineur et al., 2011). In summary, many alkaloids from the Solanaceae are likely to target the enteric nervous system in preference to the brain. Their mode of action is consistent with co-evolution of leaves and guts.

Toxins from evergreens

The production of alkaloids is not restricted to fast-growing plants such as tobacco or nightshades. Many slow-growing evergreens produce highly toxic alkaloids too. For example, only a few hundred grams of leaves from the common yew (*Taxus baccata*) leaves can kill a horse. Yew has leaves that contain an array of poisons including an alkaloid called taxine B that functions as a calcium channel antagonist to cause a variety of symptoms including abdominal pain; at higher levels, it is a potent cardiotoxin (Wilson and Hooser, 2007). Alongside yew in western Europe one often finds Holly (*Ilex aquilfolium*, Aquifoliaceae), a tree that produces theophylline, an alkaloid found in tea and one that is usually tolerated at low levels in humans. But in many animals theophylline can disrupt digestive functions and affect the heart through a variety of mechanisms including its action as an adenosine receptor antagonist. There are several reasons to suspect that alkaloids such as taxine B and theophylline, as well as molecules from other chemical classes such as grayanotoxins (which are sodium channel activity modulators made by ericaceous plants including rhododendrons), can act as defences against vertebrate herbivores in nature. The levels of these molecules are maintained in leaves over winter when insect attack is greatly reduced. This is the case for yew and holly and is also true for rhododendron. All three of these examples are commonly found in their natural environments intermixed with deciduous trees. As has been mentioned, evergreens become increasingly attractive as targets for vertebrate predation in winter.

Which types of organisms selected for the bulk of plant chemistry? Chemical clines

Many alkaloids can affect both invertebrates and vertebrates, binding to receptors, ion channels, and digestive or metabolic enzymes that are conserved in both these animal lineages.

For instance, one of several elegant studies by Ian Baldwin that has shed light on the roles of nicotine against herbivorous insects was conducted with the wild tobacco *Nicotiana attenuata* in North America. When the ability of this plant to produce nicotine (Figure 4.5) was reduced, its leaves became greatly more susceptible to attack by grasshoppers (*Trimerotropis* spp.) and beet armyworms (*Spodoptera exigua*, Lepidoptera) (Steppuhn et al., 2004). But since nicotine is an acetylcholine receptor agonist and these receptors occur in both invertebrates and vertebrates, the molecule is also likely to protect tobacco plants from the latter. What role, if any, have vertebrate herbivores—relative to their invertebrate counterparts—played in modelling the overall palette of defence chemicals found across the plant kingdom? One approach to this question involves using the faunal history of islands.

Invertebrate herbivores and microbial pathogens are found in nearly all habitats where plants reproduce, but a number of plant families have speciated in the absence of sustained pressure from vertebrate leaf eaters. Nevertheless, despite the historical paucity of these folivores on New Caledonia, the island's endemic flora has many examples of plants that use defence chemistry, and even the ancient angiosperm *Amborella trichopoda* (Amborellaceae), a plant from a family unique to the island, emits the strong smell of methyl salicylate when its bark is scratched. The New Caledonian flora can therefore be used in an effort to determine whether the historical near absence of vertebrate folivores is coupled to a reduced level of alkaloids compared with plants from other regions where vertebrate leaf eaters evolved with the plants. To do this, one can make use of the fact that surveys have already revealed that the distribution of certain chemicals in the plant kingdom is non-random and correlates with latitude. Such a survey (not including New Caledonia) was carried out years ago by D.A. Levin who discovered that the closer one is to the equator, the higher the proportion of alkaloid-producing plants to alkaloid non-producing plants. That is, there is a latitude cline in the distribution of these compounds (Levin, 1976).

In his study, Levin surveyed countries where large vertebrate herbivores were present long before the arrival of humans. But would the proportion of alkaloids in New Caledonia where such animals arrived recently with humans also fit the latitude cline? If vertebrates put pressures on plants that then lead to the diversification of alkaloids, one would expect fewer alkaloid-containing plants than are seen in the countries included in Levin's study. To do this quantitatively was not easy without the helpful collaboration of Vincent Dumontet who was working on the island to determine alkaloid levels in its flora. Of the 3000 or so native plant species on the island, about 1404 had been tested for alkaloids. Of these, 22% tested positive (Sévenet and Pusset, 1996; Vincent Dumontet, personal communication), making a rough comparison with the levels of alkaloids in other floras possible. To our surprise, the proportion of alkaloids in the New Caledonian flora fitted well on Levin's original latitude cline (Figure 4.6). The proportion of alkaloid-containing plants on the island was to be expected from its latitude, not from its faunal history.

The broader idea is that if an island has had a unique faunal history this should be reflected in the plant defence mechanisms found on it. A survey of leaf total phenolics in plants from a number of floras put their concentration ranges between 1% and 6% of plant dry weight. Near the top of this range was New Zealand where the average total phenolic content of the leaves

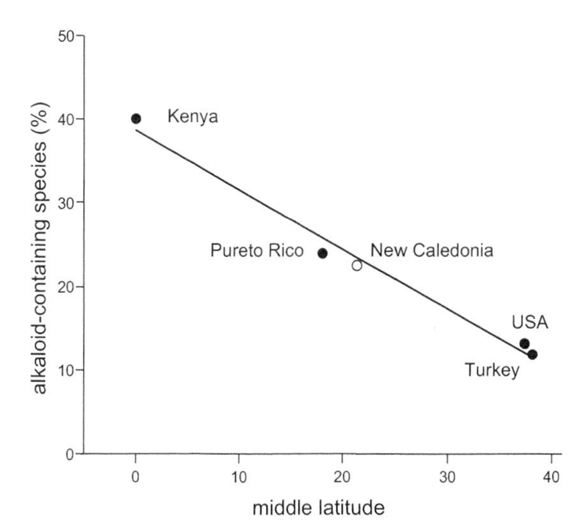

Figure 4.6 Levin's cline for the proportion of alkaloid-containing plants at various latitudes. The percentage of plants that contain readily detectable levels of alkaloids (filled circles) correlates negatively with the middle latitudes of various floras (Levin, 1976). Here, a datum for the island of New Caledonia is added (open circle); this fits the latitude cline which is significant because there is little evidence for the historical presence of vertebrate folivores on New Caledonia. The observation is consistent with invertebrate herbivores and pathogens exerting the major selection pressure for alkaloid occurrence.

of 46 species was 5.4%. Given that there is no evidence that, until the arrival of humans, the islands ever had mammalian folivores, this suggests that other organisms have selected for high phenolic levels (Bee et al., 2011). In this case, it could instead have been birds, the now extinct moas, but this seems unlikely because there are also plenty of tannin-rich leaves on plants in New Caledonia where similar, large leaf-eating birds appear to have been absent. By extending these sorts of surveys to many islands it should be possible to whittle down the number of suspects in the search for what types of organism are selected for specific types of plant defence. What we know is consistent with the idea that the presence of folivorous vertebrates may not be a major factor in determining the overall abundance of alkaloids or tannins in a flora: invertebrates and pathogens must take the blame. Instead, vertebrates have probably applied additional (sometimes strong) pressures, and compounds that evolved in the first place to target insects have been co-opted and perhaps modified to protect against vertebrates too.

Since Levin's discovery of a latitude cline for alkaloids, other attempts have been made to find similar phenomena for other classes of chemical in plants. But the current perspective is that, while there is increasing evidence supporting a latitude cline for a class of triterpenes called cardenolides (Rasmann and Agrawal, 2011), there is little evidence for this tendency in other groups of chemicals (Moles et al., 2011). However, when it comes to rates of herbivory or to overall species richness in both plants and herbivores, there is a wealth of evidence for latitude clines (Schemske et al., 2009). But the simple fact that the density of vegetation can differ greatly even at the same latitude may obliterate many defence-related clines. For example, if one follows the equator in Africa from east to west one crosses through dry parts of Kenya and Somalia with very low plant densities to the forests of Gabon with some of the highest densities of plant mass per unit of surface area on earth. Along this line there might be trends in plant chemical levels that are more closely related to plant density than to latitude.

Phenolics

Defences under nitrogen limitation: the Venus flytrap

Having looked briefly at nitrogen-based defences one can ask how plants that live in nitrogen-limiting habitats defend themselves? A good starting place to investigate this question is in families that grow naturally under nitrogen-limiting circumstances, for instance the Venus flytrap (*Dionaea muscipula*, Droseraceae). The leaves of this plant have to face a dilemma. In contrast to the leaves of most plants that will continue to function even if badly damaged, the lobes of *Dionaea* will not work as traps if they have holes in them or if the sensory hairs on the lobes are damaged. So this plant has to both attract insects and to make sure that it is not so attractive as to be eaten. Part of the nitrogen-starved flytrap's strategy to attract prey is to colour the upper surfaces of its trap lobes red. The problem is what happens when herbivores are attracted? What defence chemistry does the Venus flytrap use?

Not surprisingly, the flytrap does not rely on nitrogen-containing defence chemicals. Instead it uses one of the two major classes of nitrogen-lacking compounds: phenolics. Several flavonol glycosides (tricyclic phenols attached to glucose or galactose) are detectable if the entire trap is macerated and extracted. Additionally, the traps produce plumbagin (Figure 4.7) and this molecule is also secreted on to the trap lobe surfaces, together with some of its parent molecule plumbagin glycoside. Although it is not known exactly what plumbagin does for *Dionaea* there is no doubt that this molecule is potentially toxic—it is marketed with a bold warning label. This toxicity may be related to the fact that pumabagin is a reactive

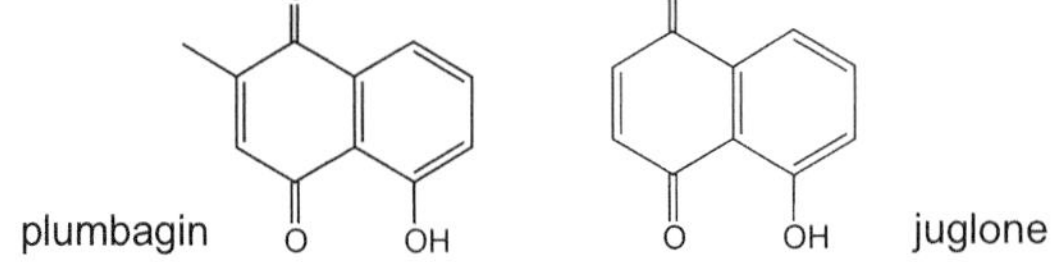

plumbagin O OH O OH juglone

Figure 4.7 Plumbagin and juglone.
Plumbagin is found on the trap surfaces of the Venus flytrap, and juglone is produced by walnut leaves. Both compounds are 1,4-naphthoquinones and are chemically reactive. In the storage forms of these compounds (not shown) the hydroxyl groups are bound to glucose.

electrophile that can bind to nucleophilic atoms in proteins (Tokunaga et al., 2004). In investigating plumbagin's role on the lobes for flytraps in the laboratory we noticed that this slightly volatile molecule can narcotize fruit flies. Perhaps it is produced to stop trapped insects struggling and thereby damaging the trap.

Juglone and walnut

Toxicity and reactivity is the case for countless other plant phenolics. Perhaps the best-known relative of plumbagin is juglone (Figure 4.7), a compound produced as juglone glucoside by the leaves and roots of walnut trees among which are the black walnut native to eastern North America (*Juglans nigra*, Juglandaceae) and the 'European' walnut (*J. regia*). Being reactive, compounds such as juglone could, in theory, harm any pathogen or herbivore, but juglone from walnut trees is best known for a different reason. These trees are reported to kill neighbouring plants through using juglone as a toxic (allelopathic) agent. Observations of this go back to Pliny the Elder whose *Natural History* summarized the first-century Roman scientific knowledge of the natural world. He wrote: 'The trees themselves and their leaves give out a poison that affects the brain' and 'The oak also dies if planted near a walnut' (Healy, 2004). Some authors have implied that Pliny wrote about today's *J. nigra* (Harborne, 1993), but, he was most likely referring to *J. regia*. This tree was well known and cultivated in Roman times whereas the majority of other walnut species (including *J. nigra*, the American walnut) grow in the New World and would have been unknown to the Romans.

What Pliny wrote about walnut trees seems to have sparked much of the later interest in *Juglans* on both sides of the Atlantic. Indeed, *J. nigra* in North America has been shown to kill-off potato, tomato and alfalfa that were grown below it (Massey, 1925). But while it is clear that juglone can be toxic to plants (Hejl et al., 1993) it is possible that the main natural role of juglone glucoside released from the European walnut *J. regia* is to repel small herbivores such as mites and/or to reduce the growth of epiphyllic organisms or pathogens. The effect of juglone when washed-off on to surrounding plants may well be secondary.

Phenolics and polyphenol oxidases

In many cases where one finds large amounts of phenolic compounds one also finds enzymes called polyphenol oxidases (PPOs). These robust enzymes usually have broad-spectrum substrate preferences using molecular oxygen to oxidize ring hydroxyl groups on phenols to produce quinones which can then polymerize spontaneously. If the activated phenols do not polymerize they can instead bind to other molecules, potentially to proteins in both pathogens and herbivores (Figure 4.8). The activity of PPOs is responsible for much of the browning of damaged plant tissues (e.g. apple fruits) and this seems to have led to the idea that these enzymes help to seal wounded tissues. But this is unlikely to be the only role of PPO: it appears to be very much defence-related; it is often wound-inducible; and it tends to have high activities in young leaves. In some cases PPOs help plants to make insect-catching glues.

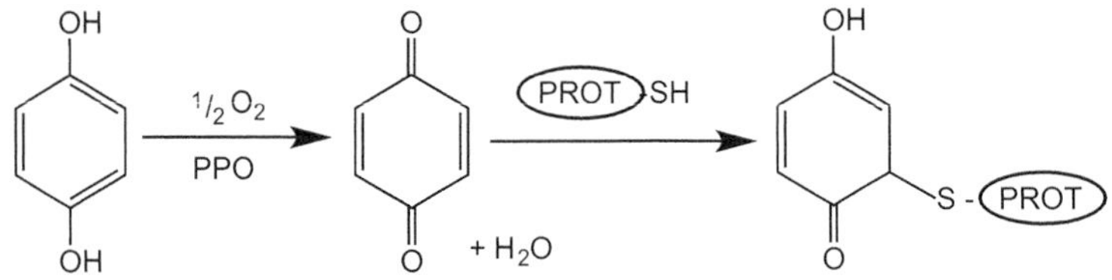

Figure 4.8 A mechanism of oxidation of phenolics by polyphenol oxidase (PPO).
Oxidized phenolics can polymerize with each other and/or they can bind to proteins. The binding to a protein thiol group is shown but amino groups can also be targets of the sorts of electrophilic phenols produced by PPO reactions.

The leaves of the wild potato *Solanum berthaultii* (Solanaceae) from western South America have both short and stubby trichomes as well as elongated trichomes. The latter contain sesquiterpenoids including the compound *E*-β-farnesene which is an alarm pheromone in some aphid species. If, by damaging an elongated trichome, an aphid triggers the release of this compound, feeding is replaced by rapid panic movement that can make matters worse since it can damage the other trichome population. As these shorter trichomes are broken by the struggling aphid they quickly release the mixture of phenolic substrates and the PPO enzyme which then catalyses the polymerization of the phenolics to form a glue that traps the alarmed insects (Gibson and Pickett, 1983; Kowalski et al., 1992; Hall et al., 2004; Constabel and Barbehenn, 2008). Whether or not catalysed by enzymes such as PPO, phenolic compounds are commonly subject to oxidation and related chemical reactions (Appel, 1993), that is, they are often reactive. One of the best places to find reactive phenols is in the plant family that contains poison ivy, the Anacardiaceae.

Poison ivy phenolics

Some of the more spectacular defences in plants must surely be in the Anacardiaceae, the cashew family. Among the most remarkable are the false mahogany trees *Semecarpus atra* and *S. neocaledonica* of New Caledonia. *S. atra* can exceed 25 m in height and, despite the fact that its lowest leaves may be out of reach of humans, each tree is has to be approached with care. A local French word for these trees is 'goudronnier', meaning bitume maker, and they constantly produce a liquid that drips on to the ground or on to the trunk, turning it black and making it look as if it has been daubed with tar. Even more so than the related poisonwood tree (*Metopium toxiferum*) found in Florida and throughout the West Indies (trees that are often marked with warning signs due to the reactive phenolics they produce), the goudronniers in New Caledonia have to be approached with great caution. The danger is to the eyes. At least this is not generally the case for another group of Anacardiaceae that are known only too well to people in large parts of North America.

Poison ivy (*Toxicodendron radicans*) and poison oaks (*T. diverisilobum* and *T. pubescens*) and some of their relatives produce urushiols, molecules that have a phenolic head group

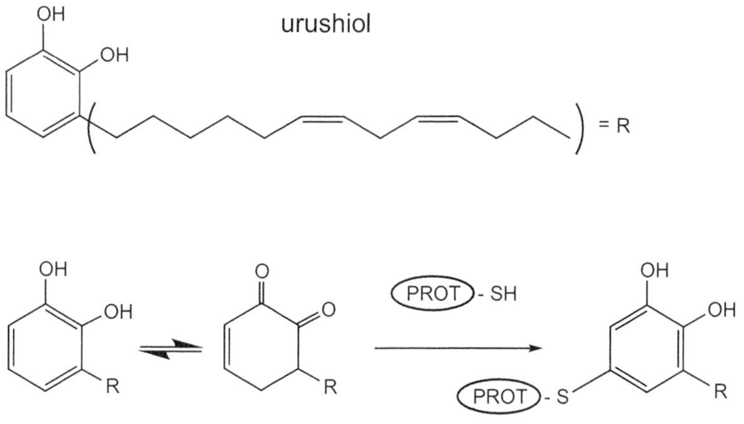

Figure 4.9 An urushiol.
Urushiols produced by poison ivy are sufficiently hydrophobic to penetrate deep into skin where they may react with proteins to create non-self molecules that are recognized by the immune system. The fatty acid side chain is variable (the example shown has two double bonds); the more unsaturated it is, the more allergenic the urushiol. A likely explanation for this is that the side chain is oxidized in human skin. However, the natural function of urushiols is likely to be independent of allergy production and is more likely to be based on binding to digestive proteins as is shown in the lower part of the figure.

and a long lipophilic tail. Being lipophilic makes these compounds highly penetrant in human skin where, from the epidermis, they migrate deep into the dermis to react with proteins (Kalergis et al., 1997) (Figure 4.9). Through this process skin proteins are converted into non-self molecules that can then stimulate an immune response. Poison ivy is usually inoffensive at the time of first contact, but by the time antibody populations have built up, a second contact with *Toxicodendron* leaves can trigger strong allergies. Urushiols volatilized from these plants by forest fires can also cause allergic reactions in the lungs. Of course triggering allergies and/or dermatitis in humans is not the primary function of these phenolics or of other reactive defence molecules. Instead, the functions of these chemicals may be to react with and inactivate proteins in the mouths and digestive systems of herbivores, or to migrate into the tissues of pathogens and epiphylls to disable their proteins. Other phenolics achieve their reactivity primarily through absorbing light energy.

Photoactivated toxins

When phototoxins absorb energy from light they can either pass this on to other molecules (and in particular to molecular oxygen) with damaging effects, or they become more reactive themselves so that they bind to other molecules (Downum, 1992; Berenbaum, 1995). The vast majority of these compounds in nature go unnoticed by humans, although this is not the case for the giant hogweed (*Heracleum mantegazzianum*, Apiaceae), a plant that produces the furanocoumarin 8-methoxypsoralen (Figure 4.10). Simple contact with leaves or stems is

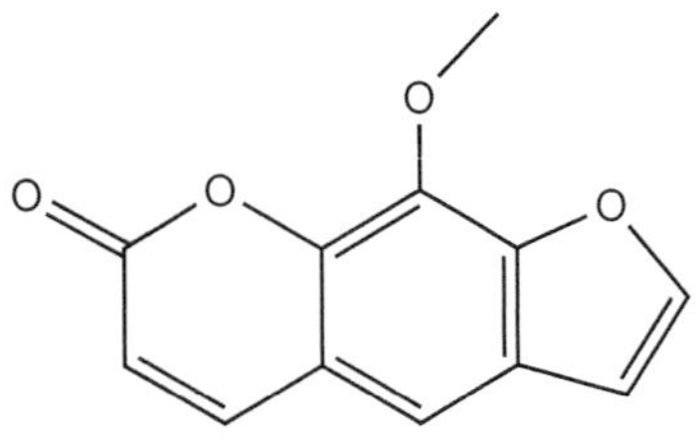

8-methoxypsoralen

Figure 4.10 8-Methoxypsoralen.
This molecule is a phenolic phototoxin from the giant hogweed (*Heracleum mantegazzianum*; Apiaceae) and some of its relatives. Phototoxins are activated by sunlight and then either react directly with other molecules (e.g. DNA) or they pass on energy to other molecules (e.g. O_2), making these other molecules reactive.

enough to cause dermatitis that is powerfully aggravated in the sunlight which activates the molecule, increasing its ability to bind to and damage macromolecules including DNA. Another example is the common European weed St John's wort (*Hypericum perforatum*, Hypericaceae) that stores the phototoxin hypericin (an octacyclic polyol) in its leaves. In the light, hypericin passes on energy to oxygen to produce singlet oxygen (Theodossiou et al., 2009), an extremely damaging reactive oxygen species that inserts itself into the nearest molecule, altering it irreversibly. Phototoxins are powerful defences against invertebrates. Again, they underscore the fact that the action of numerous defence chemicals is reactivity-based.

Oleuropein: a final example of reactivity

A final phenolic chemical that seems to distil properties of reactivity is oleuropein (Figure 4.11) from the Oleaceae family and found in the leaves of olive (*Olea europea*) and privet (*Ligustrum obtusifolium*). Oleuropein is a double-edged sword, a hybrid defence molecule build with two different components, one a phenolic unit, the other a monoterpene derivative called an iridoid glycoside. The interesting thing is that each of the two halves of oleuropein can be activated so that it can bind to nucleophilic atoms on proteins (Konno et al., 1999). The phenolic unit can be oxidized by polyphenol oxidase to a reactive quinone and the sugar residue attached to the iridoid is readily cleaved off by glucosidases. As in countless other examples of plant defence chemicals, once the sugar is removed the molecule rearranges and, in this case, a five-carbon structure with two reactive aldehyde groups is exposed. Each 'side' of oleuropein can then react with proteins, so it is a threat to leaf eaters (Konno et al., 1999).

Oleuropein has been cited here as the final example of a reactive phenolic to emphasize the contrast with a very large and common group of phenolics that have a markedly lower chemical reactivty, that do not act as phototoxins, and that are generally of low toxicity to humans. These are tannins, enigmatic compounds without well-defined roles and clearly defined receptors.

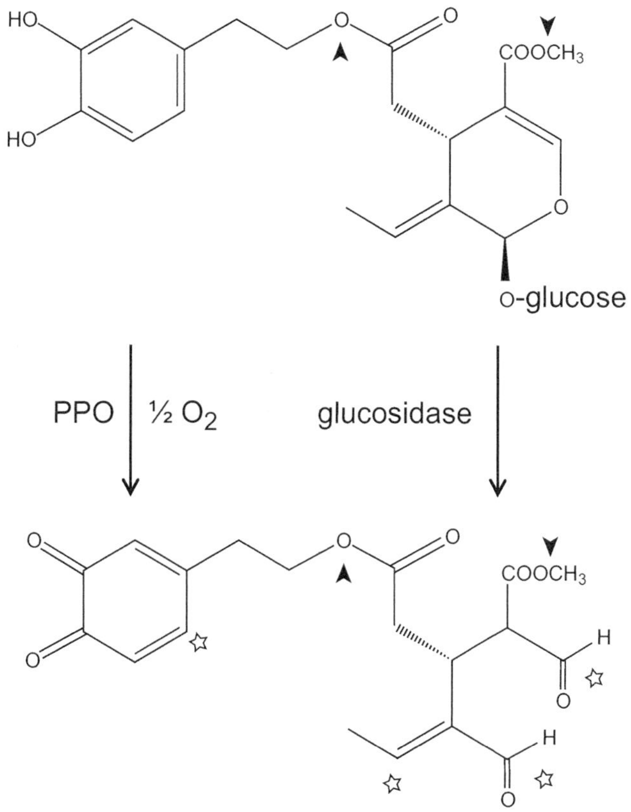

Figure 4.11 Oleuropein glucoside, a pro-reactive phenolic in the leaves of privet (*Ligustrum* spp.) and olive (*Olea* spp.).
Oleuropein exemplifies plant chemicals that are reactive and that can potentially bind non-specifically to multiple proteins. The left-hand end of the molecule can be oxidized by polyphenoloxidase (PPO) to a reactive quinone. β-Glucosidase-catalysed removal of the glucose on the right-hand side of the molecule causes the generation of two aldehyde groups. Other sites that are prone to hydrolysis are marked with black arrowheads. In the activated molecule shown at the bottom of the figure the main sites that can bind to nucleophilic atoms (N and S) in proteins are indicated with stars.

The tannin superfamily

Tannins are phenolic molecules that have long been classified into two major groups: hydrolysable tannins and condensed tannins (Figure 4.12) although there are intermediate forms (Haslam, 1998). How these compounds function is unknown, but we do know that their presence correlates with increased defence capacity of leaves. For example, the leaves of pedunculate oak (*Quercus robur*) make tannins of both class and, correlating with decreased susceptibility to the larvae of winter moths, the levels of condensed tannins increase as summer progresses (Feeny, 1970). Tannins occur widely and they often fill the vacuoles of large photosynthetic cells in leaves (Figure 4.13).

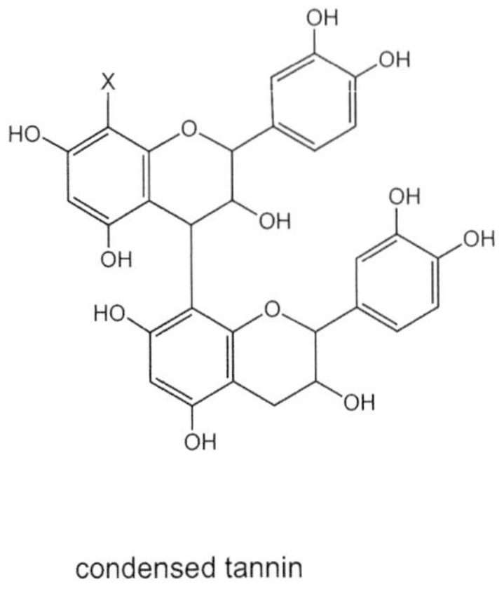

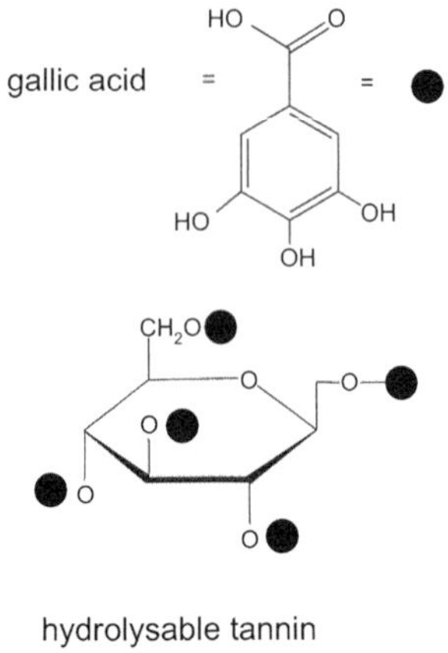

condensed tannin

hydrolysable tannin

Figure 4.12 Tannins.
Condensed tannins are based on a few variants of hydroxyflavan-3-ol and these have a 6:3:6 structure wherein the central ring is composed of three carbons and one oxygen, and where the outer rings both have six carbons. These molecules resist hydrolysis. By contrast, hydrolysable tannins are hexoses (typically glucose) decorated with phenolic units one of which is gallic acid, a seven-carbon molecule indicated as black-filled circles. This is often replaced by a more complex 14-carbon dimeric product of gallic acid oxidation called ellagic acid (not shown). X indicates that only part of a much larger molecule is illustrated.

Hydrolysable tannins: phenol-coated sugars

Hydrolysable tannins are usually composed of more than one phenolic acid unit bound to glucose. The sugar acts as the central part of the scaffold and can be bound directly by up to five gallic acids. These resulting molecules (e.g. pentagalloyl glucose, Figure 4.12) can then be decorated further with additional phenolic acids, making the molecules more complex still. To complicate things, hydrolysable tannins are not always found as molecules containing a single sugar residue. Higher complexity forms are known where there might be at least three glucose residues and up to 15 or so phenolic acids. However complex these molecules are, they have shared features. Each contains a sugar residue at its centre and, because of this, each is relatively labile. Hydrolysable tannins would seem to be made to be broken down in the digestive systems of herbivores.

Condensed tannins: phenol-linked phenols

Whereas hydrolysable tannins are phenols linked to sugars, condensed tannins are phenols linked to phenols (Figure 4.12). These molecules are very resistant to hydrolysis since they are held together by carbon–carbon bonds that can only be depolymerized at very low pHs that are rarely found in the realm of biology. So, unlike their hydrolysable counterparts, condensed tannins are built to last. Often major components of wood, they are also common in leaves and they have been repeatedly linked to the deterrence of insect herbivores. Furthermore, condensed tannins are often found in specialized defence cells.

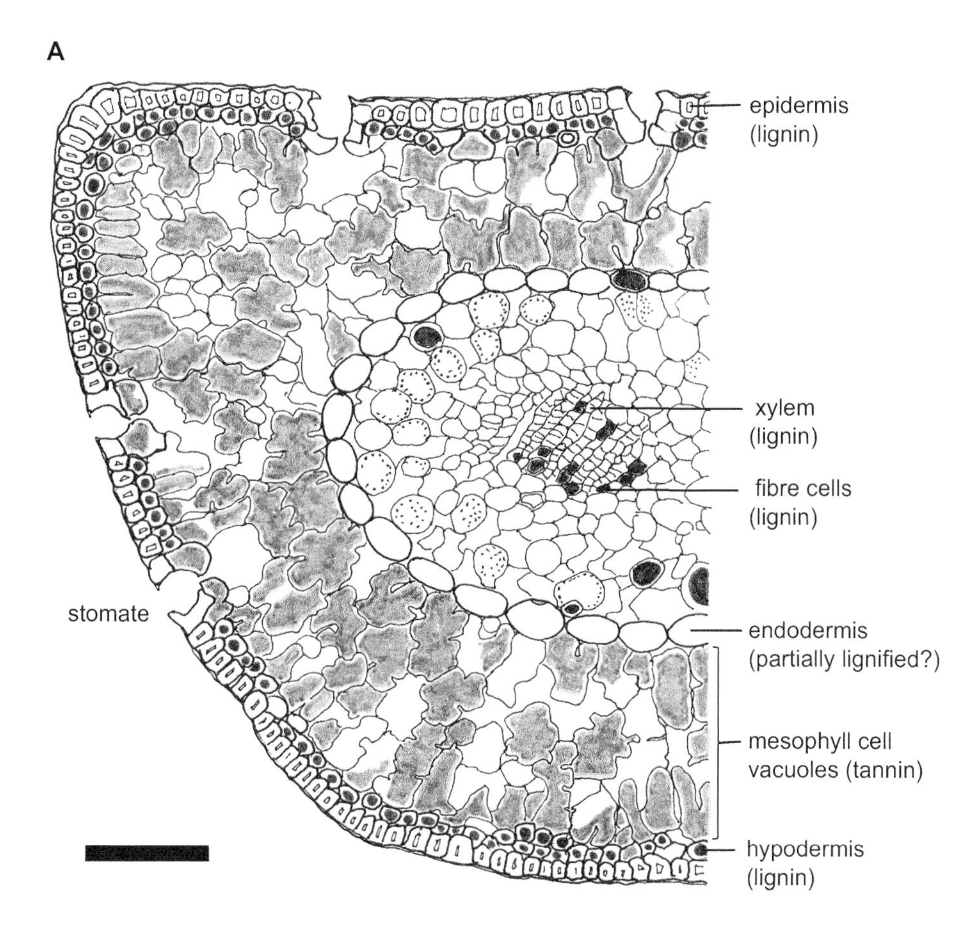

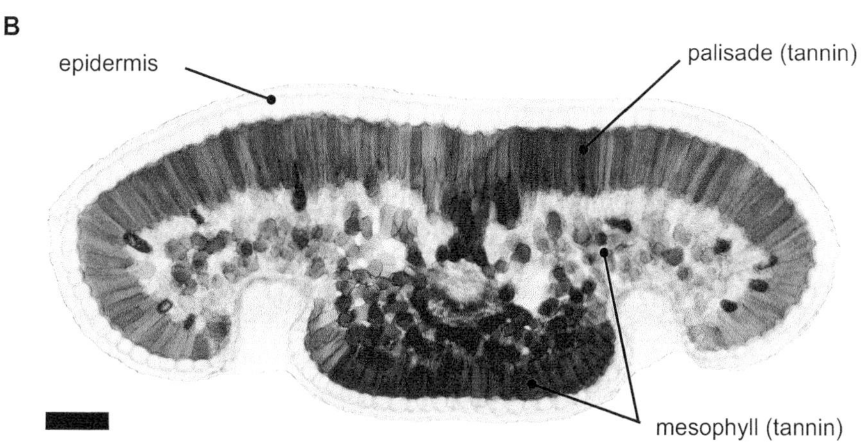

Figure 4.13 Polymerized phenolics as structural and defence compounds in two Mediterranean plants. (A) Transversal section through a needle of the Mediterranean pine *Pinus nigra*. Phenolics occur as lignin in the epidermis, hypodermis (sclerenchyma), xylem, fibre cells in the vascular bundle, and probably in the endodermis. The vacuoles of mesophyll cells contain high levels of tannins. Scale bar = 100 μm. (B) Transversal section through the leaf of Mediterranean heather (*Erica carnea*). Photosynthetic cells with high vacuolar tannin levels are indicated. Scale bar = 300 μm.

For example, in the leaves clover *Trifolium repens* (Fabaceae) and birdsfoot trefoil *Lotus corniculatus* (Fabaceae) they are sequestered in idioblasts which are most numerous at leaf bases. This strongly suggests roles in defence and some of the best candidates for their targets are extracellular proteins produced by herbivores. This assumption rests on a limited number of facts.

Tannin-binding proteins from herbivores

Tannins get their name from the ability to tan leather, that is to bind to and stabilize dominant skin proteins such as collagens. It is the fact that some of the best tannin-binding proteins are collagens that gives us hints as to which sorts of proteins might be targeted by these molecules in nature. Since a herbivore eating a leaf is more likely to get leaf chemicals in its mouth than on its skin, one should first consider salivary proteins as being potential targets for tannins. Moreover, since collagens are proline- and hydroxyproline-rich proteins, it makes sense to ask if these types of proteins exist in the saliva. They do.

The mule deer (*Odocoileus hemionus*), a browser found in western North America, produces abundant proline- and glycine-rich salivary proteins; an attractive hypothesis is that these proteins might function to precipitate tannins (Mehansho et al., 1987; Austin et al., 1989). By using these proteins to precipitate tannin, the deer might be able to recover higher quantities of essential amino acids as well as those amino acids that, unlike glycine or proline, contain more than one nitrogen atom. It is also noteworthy that some of the most abundant human salivary proteins are proline-rich and can bind tannin (Yan and Bennick, 1995).

Tannins as orphan defences

Another proposed role for tannins is in iron chelation (Lopes et al., 1999). Iron is an essential element but too much can promote damaging oxidative reactions in cells. So intracellular iron levels are kept to a minimum. It would make sense that this could act as a defence against microbes, but would it be perceived quickly enough by vertebrate herbivores to affect their feeding preferences? Until more of their true molecular targets are found, tannins are 'orphan' defences lacking well-defined tasks in defending the plant. This is of interest because not all defence chemicals need to be targeted to specific proteins (e.g. regulatory ion channels) in the herbivore. Here, a key point has to be made. Compounds in the diet can have front-end and tail-end effects on herbivores. Until now the front-end effects of plant chemicals have been discussed almost exclusively, but, before finishing the chapter with a section on terpenes, it is necessary to consider the tail-end effects of plant chemicals. Tannins will re-enter this story, after more reactive defence chemicals.

Detoxification: the important tail end of the defence process

The targeting of discrete and critically important receptors and ion channels in animals can be seen as primary, front-end effects of defence chemicals. But there is also a tail end of chemical defence: the process of detoxification. The tail-end effects of chemicals—the fact that they have to be detoxified—can be as important as their front-end effects. The sorts of molecules that are often produced in large quantities in leaves (molecules such as monoterpenes and many types of phenolic compound including tannins) appear to have more modest effects than a poisonous alkaloid such as strychnine and in most cases they do not yet have clearly identified targets in animals. Nevertheless, by exploiting tail-end effects and forcing animals to expend energy in the detoxification process, a plant can fully redeem the costs of making the chemicals. Whatever the site in the herbivore that is targeted, a co-evolved animal will somehow have to avoid the defence molecule, destroy it, evacuate it, or sequester it. Chemical features of the defence compound other than those conferring its ability to fit exactly into a binding site in a receptor can now come into play. To balance the investment in chemical defences the leaf could, in theory, make chemicals that force leaf eaters to detoxify them in costly ways.

One of the most widespread and most frequently employed detoxification mechanisms in both invertebrates and vertebrates involves first oxygenating the defence chemical. The strategy used is simple: to make the molecule soluble to facilitate its excretion. The cytochrome *P*450 enzymes that initiate the process are chiefly mono-oxygenases that can add a single oxygen atom into the carbon skeleton of the defence metabolite to produce a hydroxyl group. Once in place, this group can be used by the herbivore to conjugate the plant defence compound to a polar molecule such as a uronic acid, a sugar, or malonate, etc. At this point, the conjugate is polar and it can be excreted readily provided that the herbivore has the carbon and energy to spare to complete the whole process (Schuler, 1996). In some cases this type of detoxification system has evolved extraordinary efficiency. The Australian marsupials that eat only the leaves of *Eucalyptus* have detoxification systems that do not waste carbon. Good examples of this come from the work of William Foley.

Koalas spend most of the day inactive, part of the reason being that long periods are needed for digestion and detoxification. Stasis saves the animal calories. Before feeding, koalas often shake branches and, with their stikingly large noses, smell the leaves as a means of assessing their palatability. Volatile terpenes and non-volatile formylated phloroglucinols (phenolic derivatives) act as feeding deterrents, perhaps making it more difficult for the koalas to find leaves that are rich in nitrogen (Moore et al., 2010). Even though it is so highly adapted to feeding on eucalyptus, only about one-tenth of the koala's low energy requirements come from cell wall polysaccharides—the rest comes from more readily digested cell contents such as starch and protein. But to get at these, the animals must also ingest high levels of eucalyptol and cymene and related monoterpenes. In most mammals these types of compounds would be oxidized and then conjugated to carrier molecules prior to excretion. But koalas and greater gliders, another leaf-eating marsupial, do it differently. Both of these animals, which are obligate *Eucalyptus* folivores, can heavily oxygenate the monoterpenes and they excrete them without conjugating them to carrier molecules. Ringtail possums that both consume *Eucalyptus* leaves and the leaves of other plants are intermediate in their ability to oxidize cymene and eucalyptol. But rats that have no evolutionary history in a *Eucalyptus* forest, are very poor oxidizers of the two compounds (Eschler et al., 2000; Dearing et al., 2005; Foley and

Moore, 2005; Marsh et al., 2005) and the monoterpenes simply place too much of a burden on the rat's detoxification capacities. In summary, generalist herbivores with broad diets do not have to focus on detoxifying one particular class of molecule. They move from one species of plant to another plant, in part to minimize the abundance of any one defence metabolite in their digestive system. But highly specialist folivores have to be experts in disarming and excreting the chemical defences of the leaves they concentrate on.

Two hypothetical plant strategies to place a burden on detoxification

Returning to the phenolics and using them as examples, one can now see two main ways in which the production of high levels of non-specific chemicals (that is, those apparently lacking a single important molecular target in the host) might burden an organism's detoxification system. The first of these relates to reactive plant chemicals, the second to less reactive molecules exemplified here by hydrolysable tannins.

Often, when there is conflict, as there is between plants and herbivores, an arms race leads to the production of high levels of defences. This has occurred in the leaves of countless plants, and in many cases where large levels of low-molecular-mass defences are made these molecules are chemically reactive. In the plant they are stored in less active forms (e.g. as glucose conjugates), but, once released into the mouth of a herbivore their reactivity is exposed. Furthermore, many of these chemicals share common features and they are likely to be detoxified in much the same way by any animal that tries to eat the leaf. Most of these chemicals contain no nitrogen, and they are relatively cheap to make—but they may cost the leaf eater far more to detoxify.

Chemically reactive compounds abound in the plant kingdom. They range from 1,4-naphthoquinones, urushiols, isothiocyanates released from glucosinolates, and many terpenes. There are so many of these molecules that it is worthwhile searching for general patterns in the ways they might work to the detriment of herbivores. The example to be taken here is the large number of plant-derived defence compounds that contain molecular motifs known as α,β-unsaturated carbonyl groups (Figure 4.14). Examples of these structures are found in 1,4-naphthoquinones such as plumbagin and juglone, and many other phenolics including oleuropein. An equally large number of molecules are proelectrophiles. These can be converted into chemicals containing the α,β-unsaturated carbonyl motifs in or on the herbivore, as is the case in urushiols. Simple chemical features of molecules, such as aldehydes or α,β-unsaturated carbonyl motifs, can react with major cell protectants such as glutathione or with sulphur and nitrogen atoms in proteins, so the presence of chemicals containing these structures can be detrimental to organisms that attack plants (Farmer and Davoine, 2007). Such molecules represent a threat to cells because they are constantly generated through non-enzymatic lipid oxidation. Cells are therefore well equipped to deal with them.

A major route to the deactivation of α,β-unsaturated carbonyl compounds is to reduce the α,β double bonds at the expense of cellular reductants (usually reduced nicotinamides) in the animal. If leaf defence chemicals could cause the depletion of reductants in the herbivore's cells this would compromise its attacker. Everything suggests that this does indeed

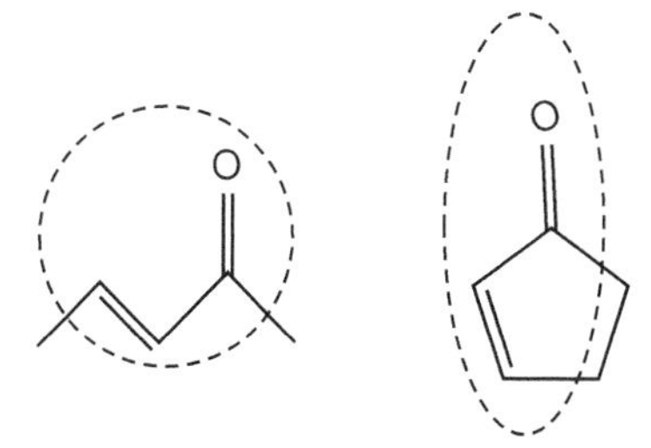

Figure 4.14 The α,β-unsaturated carbonyl group, a common motif in plant defence chemicals. All organisms are well adapted to cope with this widespread and chemically reactive molecular motif. Once inside cells, the double bond adjacent to the carbonyl group in molecules containing this structure is reduced at the expense of nicotinamide dinucleotides which are important cellular reductants. By exposing herbivore digestive systems to high levels of compounds containing this motif, leaves can place a burden on the animal that will then need to consume additional energy sources to restore the reduced form of the dinucleotide.

happen. Give any one of these compounds to a crude extract of an animal (or plant) and, provided that the adequate co-factors such as nicotinamides are available, the compound will be reduced rapidly by abundant oxidoreductase enzymes. Therefore reactive electrophiles are problematic to cells that are exposed to them. If they enter the cells they will be quickly reduced, or, alternatively, they will be coupled to glutathione, usually with the aid of glutathione-*S*-tranferase enzymes. Enzymes intercept exogenous reactive compounds so aggressively that the herbivore is essentially 'forced' to choose between having its valuable proteins and amino acids modified by the chemical, or expending resources to detoxify the compound. Just how much this type of 'reductant draining' mechanism really adds to the burden on the folivore is not yet clear but there are growing indications that herbivores try to diminish the burden imposed by electrophiles. For example, work on fatty acid-derived compounds containing α,β-unsaturated carbonyl groups indicates that enzymes in the guts of specialized herbivores can target this double bond and can isomerize it to a position in which the motif has greatly reduced chemical reactivity (Dabrowska et al., 2009). But in the case of molecules such as oxidized urushiol the reactive motif is positioned in the parent molecule in such a way that its double bond is not readily disarmed by isomerization.

Finally, we are left with tannins and in particular with hydrolysable tannins—molecules that lose their tannic properties when they are hydrolysed. How might these work in defence? It seems possible that they act as tail-end defences because, like α,β-unsaturated carbonyl compounds, they burden detoxification. But in this case the mechanism would be different. Gallic acid (Figure 4.12), a robust and polar molecule, is perhaps a pre-made substrate for excretion. All it needs is to be ingested and to be transported through the gut wall to enter an animal's circulation. If it did, it would likely be conjugated to a small molecule, a sugar or a malonate group, for example, and then it could be excreted again, at a cost (since carbon is cheaper for the plant than it is for the animal). It should be possible to test this hypothesis once we can engineer tannin production.

Terpenes

The last of the main classes of plant chemicals is also the largest and most complex. These are the terpenes (Gershenzon and Dudareva, 2007). As already mentioned, polymerization is common in the phenolics. But it is also very characteristic of terpenes—all terpenes derive from highly controlled polymerization reactions. The basic isoprene building block used to build terpenes has five carbons and the smallest of the terpene defence compounds, monoterpenes, have 10 carbons. Sesqui-, di-, and triterpenes contain 15, 20, and 30 carbon atoms respectively. However, during their synthesis, secondary reactions often modify the number of these atoms, so it is not uncommon to find 14-carbon sesquiterpenes or 27-carbon triterpenes, etc. Furthermore, terpenes are often modified by oxygenation and glycosylation (sugar addition) and many exist as multiple isomeric forms. This is where the extraordinary complexity of the terpenes comes from. This chemical class contains some of nature's most potent biocides such as pyrethrins, insect sodium channel poisons from the flowers and leaves of some *Chrysanthemum* (Asteraceae) species. To look at examples of defence-related terpenes one can begin with monoterpenes, then progress step by step to giant polymerized terpenes. This encompasses two of the best-known defence substances, resin and latex—mixtures that differ quite fundamentally in their physicochemical properties and yet are built in large part from the same five-carbon precursors from which all terpenes derive.

Monoterpenes

Beginning with monoterpenes, the leaves of mints and peppermints (*Mentha* spp; Lamiaceae) are scented with oxygenated monocyclic terpenes such as menthol whereas the West Indian lemon grass *Cymbopogon citratus* (Poaceae) owes its smell largely to the presence of linear monoterpene aldehydes such as citral. Being the smallest members of the terpene family, monoterpenes are in some respects also the easiest to study. For example, it has been possible to engineer the production of a monoterpene alcohol called linalool into the leaves of *Arabidopsis thaliana*, a plant that normally only produces trace quantities of this substance in its flowers. The transgenic plants repelled aphids far more effectively than the control plants that did not produce linalool in their leaves, providing direct evidence that monoterpenes can act as insect repellents (Aharoni et al., 2003; Gershenzon and Dudareva, 2007). Many other monoterpenes have similar roles but they are not always clear.

Of great interest to most domestic cats is nepetalactone (Figure 4.15), a bicyclic monoterpene present in the leaves of catnip (*Nepeta cataria*, Lamiaceae). Nepetalactone illustrates the fact that chance combinations of plants and animals can reveal properties of molecules that might have no ecological relevance. Indeed, our own propensity to ascribe function to the more-or-less random effects of various plant chemicals on certain animals can be termed the 'catnip syndrome'. But ecological relevance is just what many monoterpenes have. After all, these are major defence chemicals in dominant plants throughout the Mediterranean basin, in the extensive eucalyptus forests of Australia, as well as in conifers in the great northern forests, these being but a few examples.

Plate 1 Herbivores can manipulate leaf physiology. Leaf-mining insects risk being isolated in senescing leaf tissue or being shed with leaves in the autumn. As a strategy to complete their life cycles before this happens, many miners either delay leaf fall and cause the formation of green islands in which leaf tissue remains healthy while the rest of the leaf senesces. The leaf shown is from hornbeam (*Carpinus betulus*, Betulaceae) and was probably parasitized by the larva of a pygmy moth (*Stigmella* sp., Lepidoptera). (See Figure 1.4.)

Plate 2 Tolerance to damage, a quintessential feature of leaves. This leaf of *Tussilago farfara* (coltsfoot, Asteraceae) from the European Alps has been severely damaged by chrysomelid beetles but is still partly functional. Leaves show more tolerance to damage than most other organs in nature. The leaf was ~20 cm in width. (See Figure 1.6.)

Plate 3 Dark leaf flecks as putative warning signals. *Kalanchoe marmorata* (about 25 cm in height) growing on the Liban plateau, Ethiopia. (See Figure 2.2.)

Plate 4 Spotted leaves on orchids may serve as warnings. *Dactylorchis latifolia* growing in Fribourg, Switzerland. The dark purple leaf markings are displayed only on the upper leaf surface. They are present well before flowering is initiated and, upon the initiation of production of the floral stem, they often fade. Scale bar = 2 cm. (See Figure 2.3.)

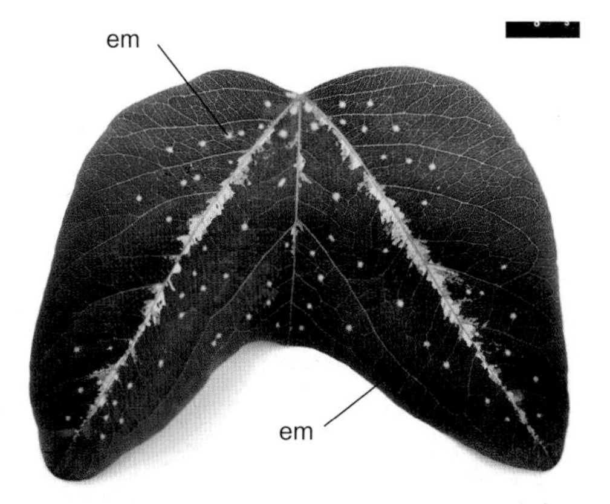

Plate 5 Deceptive patterns on passionflower leaves. To deter egg-laying by heliconid butterflies the leaves of *Passiflora boenderi* from Costa Rica have yellow egg mimic (em) spots on their upper surface. Not shown is the underside of the leaf. Below the egg mimic spots are extrafloral nectaries that produce sugar to attract protective ants. (See Figure 2.5.)

Plate 6 Stinging nettles from the southern hemisphere. (A) Leaf of the Gympie nettle *Dendrocnide moroides* from Queensland, Australia, that is known for causing long-duration pain. Scale bar = 1 cm. (B) Ongaonga (*Urtica ferox*) from New Zealand has a hard-hitting, fast-acting sting. Scale bar = 2 cm. (C) *U. ferox* showing detail of sting. Scale bar = 8 mm. (See Figure 3.4.)

Plate 7 An example of sand armour. Sand armour on the desert plant *Indigofera argentea* (Fabaceae), Jizan, Saudi Arabia. Stems, leaf edges, and the undersides of leaves are covered in regular-spaced sand grains. Scale bar = 1 cm. (See Figure 3.6.)

Plate 8 The nickel tree, *Sebertia acuminata*. Cuts into the bark of this tree cause the release of a blue latex that contains 26% dry weight nickel. The laticifers containing this latex extend into the leaves. (See Figure 4.1.)

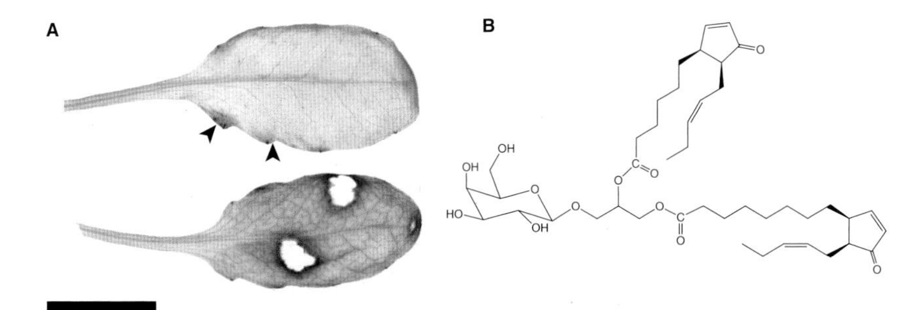

Plate 9 Highly inducible defence molecules in *Arabidopsis* leaves. Defence proteins and defence metabolites accumulate simultaneously when leaves are wounded by chewing herbivores. (A) Plants were engineered to express the promoter of the *VEGETATIVE STORAGE PROTEIN 2* (*VSP2*) gene coupled to a gene encoding a reporter enzyme (β-glucuronidase) that converts a colourless substrate to a blue precipitate. The upper leaf was from an undamaged plant whereas the lower leaf had been attacked by larvae of the Egyptian cotton leafworm (*Spodoptera littoralis*). The blue staining indicates activation of the *VSP2* gene and arrowheads show basal promoter activity in an undamaged leaf. Scale bar = 1 cm. (B) An arabidopside. These lipid derivatives are highly inducible at the edges of wounds where their concentrations can increase to >1000-fold those in undamaged leaves. Unlike VSP2, which accumulates in the timescale of hours after wounding, arabidopside accumulation begins in seconds after damage. Both VSP2 and arabidopsides have established roles in defence (Liu et al., 2005; Glauser et al., 2009). Image: S. Stolz and E. Farmer. (See Figure 5.2.)

Plate 10 Insect growth on plants lacking the ability to make jasmonates. The plant on the left is wild type whereas the plant on the right is a jasmonate biosynthesis mutant (*lox2-1 lox3B lox4A lox6A*; Chauvin et al., 2013). Neonate *Spodoptera littoralis* larvae were allowed to feed on the plants for 11 days, at which point the mass of the insects that fed on the jasmonate mutant was greater than that of the insects that had fed on the wild type. Note the damage to the central growth region of the jasmonate mutant. Despite removal of the apical meristem these plants will achieve some seed production through the activation of lateral floral meristems. Scale bar = 1 cm. (See Figure 5.4.)

Plate 11 Extrafloral nectaries on petioles in wild cherry (*Prunus avium*). (See Figure 6.1.)

Plate 12 Mite domatia on the undersides of leaves in laureltinus (*Viburnum tinum*). (See Figure 6.3.)

Plate 13 Coastal limestone habitat on north central Socotra. The pale-barked woody shrub in the foreground is *Jatropha unicostata* and this plant, ~1.5 m high, shares this habitat with caustic latex-producing spurges including *Euphorbia arbuscula* (Figure 4.18) and *E. spiralis* (Figure 8.3B). Also common in this habitat is another Euphorbiaceae, *Croton socotranus*. Just in the picture, lowest right, are the dark shoots of a severely browsed *Commiphora* (Burseraceae). The two-stemmed plant at rear left is a desert rose (*Adenium obesum*, Apocynaceae). (See Figure 8.1.)

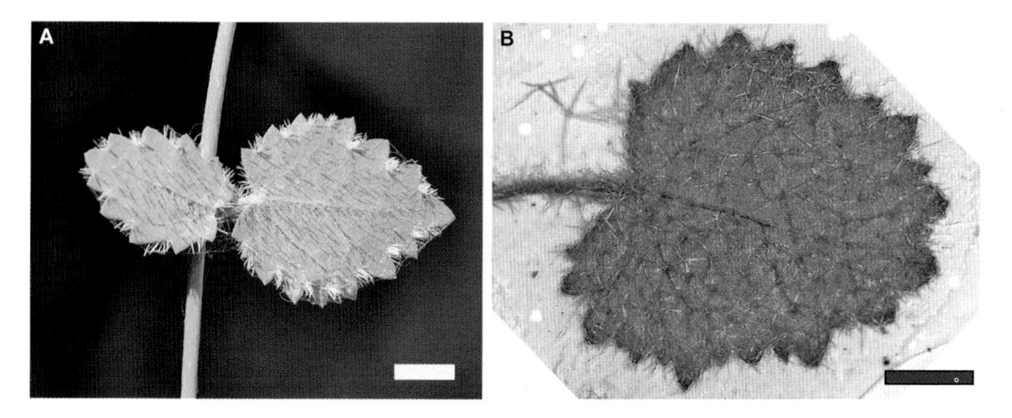

Plate 14 Stings and irritant hairs on Socotra. (A) The leaves of *Tragia balfouriana* (Euphorbiaceae) showing axial stinging hairs. Scale bar = 1 cm. (B) A leaf of *Hibiscus noli-tangere* (Malvaceae), a plant that releases highly irritant stellate hairs on contact. Scale bar = 0.2 cm. Note the similarity of leaf forms. (See Figure 8.2.)

Plate 15 Examples of thorns on endemic Socotran plants. (A) *Barleria tetracantha* (Acanthaceae). Scale bar = 0.2 cm. (B) *Euphorbia spiralis* (Euphorbiaceae). (See Figure 8.3.)

Plate 16 A cliff-dwelling Socotran *Boswellia*. The tree is *Boswellia bullata* (Burseraceae), ~5 m high. Note the massive holdfast. Photo: L. Banfield, Royal Botanic Garden Edinburgh. (See Figure 8.4.)

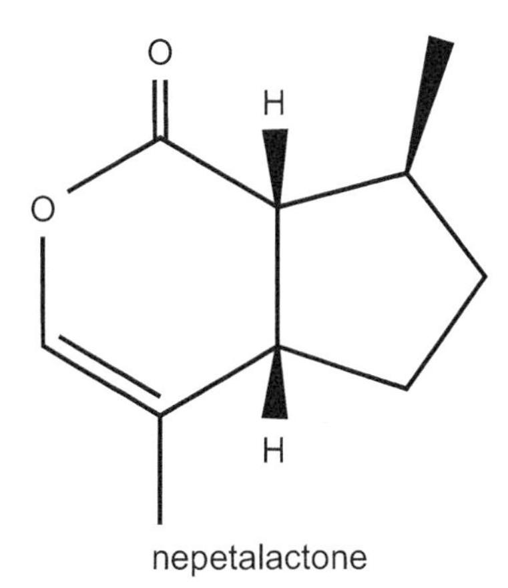

Figure 4.15 Nepetalactone from the leaves of catnip.
Nepetalactone illustrates the fact that plant chemicals sometimes elicit behavioural or developmental effects in animals, but that these effects may not always have ecological relevance. In this case the molecule elicits what appears to be euphoria in a high proportion of domestic cats.

Conifer resins: two-component defence systems

Oleoresins are resins of a lipophilic nature and are produced in trees such as firs, spruces, and pines. It is the monoterpene components of these fluids that give them much of their odour, and from a human perspective, some resins are among nature's most exquisite perfumes. Examples could be the sublime aroma of Arolla pine (*Pinus cembra*, Pinaceae) from central Europe or, in the same family, the Himalayan cedar (*Cedrus deodara*) from central Asia. However, what is agreeable to humans in minute concentration in the air is not tolerable in liquid form in the mouth and throat. Being organic solvents, oleoresins are stored outside cells in resin ducts that are arrayed so as to rupture when an organism tries to feed on a leaf. How most oleoresins work in defence is best captured by describing them as two-component systems based on petrol-like solvents made of a few types monoterpenes. Like petrol, this fluid is of low viscosity, is often flammable, and contributes most of the smell of the resin. The other major resin components, the bulk of the non-volatile portion in most conifers, are larger, non-volatile, 20-carbon diterpene acids (Trapp and Croteau, 2001).

An example of a two-component resin system is illustrated in Figure 4.16: a section through the needle of a Douglas fir (*Pseudotsuga menziesii*, Pinaceae) with two large resin ducts running along the periphery of the needle. When a herbivore attacks the leaves of firs or other conifers, it risks breaking into resin ducts and releasing their pressurized contents. Immediately, the low viscocity monoterpene solvent carries the resin deep into the mouthparts. The monoterpene solvent then evaporates quickly, leaving the non-volatile diterpene acids to coagulate into a low-pH and high-viscosity mandible-gluing fluid that begins the slow process of solidification, a process that can take hundred if not thousands of years to complete, producing amber. Of course the fact that amber often contains insects is consistent

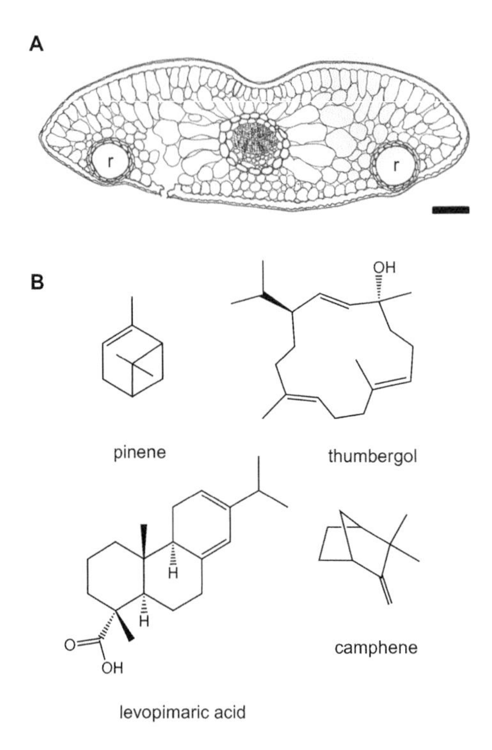

Figure 4.16 Resin chemicals and resin ducts in fir needles.
(A) Transversal section of a Douglas fir (*Pseudotsuga menziesii*) needle shows the paired lateral resin ducts (r) in which the resin is stored. Scale bar = 150 μm. (B) Four of the major terpenes found in the resin of the needles of the Douglas fir. Pinene and camphene are monoterpenes; they make up much of the fluid component of the resin. Solutes in this fluid are diterpenes such as thumbergol and levopimaric acid.

with its role in protecting trees against invertebrates, but oleoresins almost certainly protect conifer needles from vertebrates too. Conifer needles are nevertheless a winter survival food for large boreal herbivores and even some birds, although in general, they escape heavy vertebrate predation in summer. Remarkably, a few select vertebrates have adapted to counter these defences. The red tree vole (*Arborimus longicaudus*), which feeds on Douglas fir needles in western forests in the USA, bites out the central part of the needle, shedding peripheral slivers of needle that contain the resin ducts. This animal is exceptional and, in general, storing two-component organic solvent-based resins in specialized defence compartments, resin ducts, is one of the most tried-and-tested defence strategies in leaves. Another is the production of water-based emulsions called latexes.

Latexes

Latexes are opaque emulsions containing light-scattering particles of water-insoluble terpene macropolymers (1,4-polyisopropenes). Light scattering, often combined with pigmentation, gives latexes colours ranging from blue (e.g. in the nickel tree latex (Figure 4.1)) to yellow, orange, pink, and even blood red, as in some African *Macaranga* species (Euphorbiaceae). These latter latexes can be so blood-like that one's instant reaction when cutting through a stem is to think that one has cut oneself. At least 40 plant families (Agrawal and Konno, 2009) produce latex and among these are the Euphorbiaceae and the Apocynaceae—latex-producing families par excellence. Many plants in these families release latex in prodigious quantities when a plant is damaged and they often combine its production with other defence strategies. Many of these plants are spiny; some have spectacular colour patterns on their vegetative tissues.

Much of what we know about latex comes from the rubber tree *Hevea brasiliensis* (Euphorbiaceae), where this fluid is stored in articulated anastomosing lactifers below the bark and extending into the leaves. The bulk of this latex, nearly 60% by volume, is water with the major chemical component, about 35%, being particles of giant, highly polymerized terpenes: *cis*-1,4-polyisoprenes. Other minor components of *Hevea* latex include defence proteins such as endochitinases, lectin-like proteins (prohevein and pseudohevein), glucanases and also proteinase inhibitors (Chow et al., 2007). These are the allergy-inducing components in rubber. However, for a plant in the Euphorbiaceae, and viewed from a human perspective, rubber does not produce a particularly nasty latex and this greatly facilitates its harvest.

If the leaves or fruits of the aptly named apple of Sodom (*Calotropis procera*, Apocynaceae) are damaged they exude large quantities of bright white latex that can literally drown a small attacker. The latex of this plant that is encountered frequently in arid regions of India, the Middle East, and parts of Africa (and which withstands the presence of camels even in the camel market in Riyadh) carries triterpene toxins including the Na^+/K^+-ATPase inhibitor calotropin (Figure 4.17), a compound with a mode of action similar to that of digoxin and ouabain. Skin contact with the apple of Sodom's latex is not too unpleasant but some latexes cause severe chemical burns. For example, *Euphorbia poisonii* in Central Africa contains the diterpene derivative resiniferatoxin that, like capsaicin from chilis, activates the transient receptor potential vanilloid 1 (TPRV1) receptor causing a heat sensation. But resiniferatoxin does this tens to hundreds of times more potently than capsaicin (Szallasi and Blumberg, 1989), making it far too potent and damaging to use as a spice. I can attest to just how caustic some latexes can be, having unintentionally experienced skin burns with *Euphorbia arbuscula* on the Indian Ocean island of Socotra (Figure 4.18). A drop that is not wiped off immediately burns deeply and leaves a scar that is slow to heal and is painfully sensitive to sunlight.

However, latex functions as a defence not on the skin but in the mouth and throat. One way to appreciate just how irritant latex can be to generalist insects is to apply tiny amounts of it to their mandibles. If latex from Cypress spurge (*Euphorbia cyparissias*) is applied to the mouthparts of *Spodoptera littoralis* the insect instantly starts to regurgitate into its mandibles in a frantic attempt to clean them. This causes the head capsule to change colour to green and the rest of the body changes colour too (Figure 4.19). The latex from this spurge must contain potent toxins and, given the reaction of the caterpillars, these are likely to be targeted to specific receptors in the herbivore.

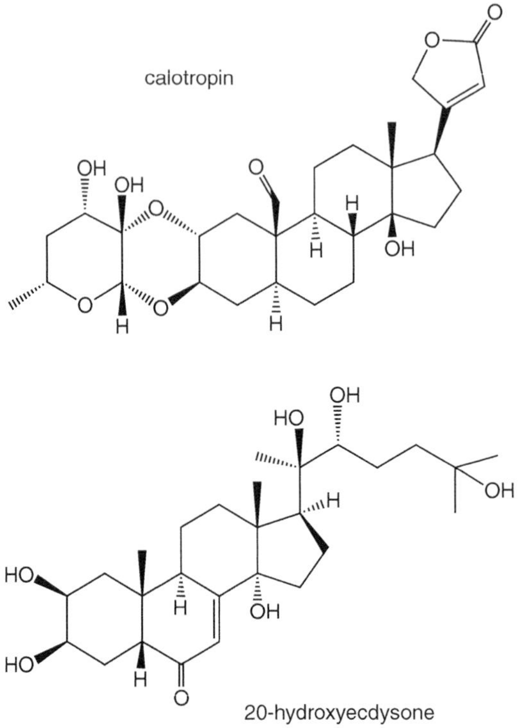

Figure 4.17 Triterpene defence compounds.
Calotropin, a cardenolide from the apple of Sodom (Apocynaceae) inhibits animal Na^+/K^+ ATPases. 20-Hydroxyecdysone produced in gymnosperms such as common yew is a ligand for ecdysone receptors in insects, and, when given exogenously to insects, can disrupt their normal development.

Cardenolides and cardenolide glycosides

Triterpenes, of which calotropin (Figure 4.17) is an example, are common defence chemicals. They are, for example, the major solutes in resins in the Dipterocarpaceae, a vast and important family of tropical plants and trees. Some, e.g. saponins, are soapy, chaotropic molecules that may have natural roles in perturbing membrane function—among these are some powerful molluscicides. But perhaps the best known of these molecules are the cardenolides, steroidal triterpenes found either as aglycones or as cardenolide glycosides, otherwise named cardiac glycosides. Cardenolides are well known from dicotyledon families, but they also occur in some monocotyledons, notably in Convollariaceae and Hyacinthaceae.

Many cardenolides, as their name suggests, are potent cardiotoxins but they also affect cells elsewhere in the bodies of both vertebrates and invertebrates. For example, several parts of the Balkan foxglove (*Digitalis lanata*, Plantaginaceae), including the leaves, contain digoxin, a cardenolide that binds to plasma membrane Na^+/K^+-ATPase in animal cells. This

Figure 4.18 *Euphorbia arbuscula* on Socotra, northwest Indian Ocean.
This tree, which can reach ~5 m in height, produces an extremely caustic skin-burning latex. Nevertheless, Socotran goats have begun the process of adapting to be able to feed on young *E. arbuscula* plants. Goats from the Arabian mainland, on the other hand, avoid the seedlings of this tree (Miller and Morris, 2004).

ATP-driven pump maintains the correct balance of sodium and potassium inside cells. Cardenolides bind the α-subunit of the pump and inhibit its function. In the case of heart pacemaker cells this leads to increased intracellular Ca^{2+} levels, increased force of heart contraction and, in too high a dose, to heart malfunction. Ouabain, a triterpene that is produced by some plants in the Apocynaceae, also targets the Na^+/K^+-ATPase but some specialist insects are resistant since they produce variant forms of the protein's α-subunit. Remarkably, similar mutations have arisen in very different types of insects ranging from monarch butterflies to some chrysomelid beetles, some hemipterans, and at least one fly. One of these mutations, for example, converts an asparagine residue into a histidine residue, lowering the affinity of ouabain for the pump and allowing these insects to feed on milkweeds (*Asclepias* spp.) free from competition from most generalist insects (Holzinger and Wink, 1996; Labeyrie and Dobler, 2004; Aardema et al., 2012; Dobler et al., 2012).

Ouabain and related molecules are potent toxins for most invertebrates. Without a resistant Na^+/K^+-ATPase the larvae of the cabbage looper (*Trichoplusia ni*), a generalist herbivore, became paralysed for up to three days after trying to feed on *Ascelpias curassavica* (Dussourd and Hoyle, 2000). However, some specialist insects employ cardenolides for their own defences against vertebrates. Monarch butterfly larvae sequester calotropin from milkweeds and retain it in an unaltered form (Nishida, 2002) as a defence against birds. The phenomenon of sequestration (particularly when a molecule from a plant is unaltered by a specialist herbivore) underlines the possibility that it might serve as a protectant against both invertebrates and vertebrates.

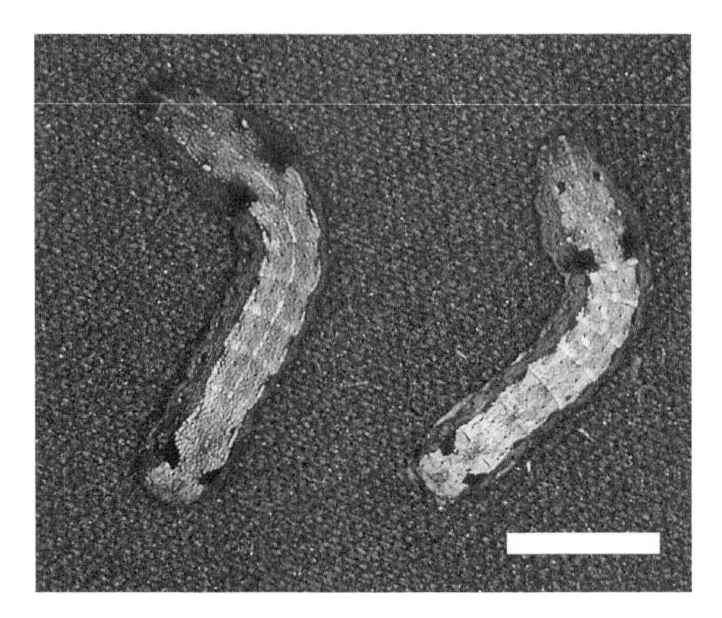

Figure 4.19 *Euphorbia* latex strongly affects insect herbivores.
Spodoptera littoralis caterpillars are generalist herbivores that react violently and can even change colour when exposed to latex. The larva on the left had 0.5 μL of water applied to its mandibles. The larva on the right was treated with 0.5 μL of latex from *Euphorbia cyparissias*. The head capsule of this larva immediately filled with regurgitant. Note the paler colour of the dorsal surface of the latex-treated insect on the right. Scale bar = 1 cm.

Triterpene hormone mimics

Hormone mimics—plant chemicals that behave like endogenous hormones in animals—occur quite widely in leaves and some of these are terpenes. For instance, some plant-produced hydroxysterol triterpene derivatives mimic the effects of ecdysones, hormones that regulate moulting in insects and other invertebrates. The moulting process is a period of vulnerability for the insect and must be executed with great efficiency so that the animal can quickly shed its old exoskelton, expand and harden its new exoskeleton, and resume its normal lifestyle. This is disrupted if insects take in ecdysones in too high a concentration or at the wrong time. Ecdysones and their plant-made mimics (phytoecdysones) are ligands for the ecdysone receptor. Therefore, if an insect ingests an excess of these molecules, it may not be able to shed its old cuticle and will become blocked in one of its larval stages, unable to reproduce, and potentially be less able to escape its predators.

The first plant found to produce phytoecdysones was the common yew, which accumulates relatively high levels of 20-hydroxyecdysone (Figure 4.17) in leaves (reviewed by Harborne, 1993). Other gymnosperm groups such as cycads produce ecdysone mimics, an example being cycasterone from *Cycas* (Cycadaceae), and similar molecules occur in the fern family Polypodiaceae, for example in the widespread *Polypodium vulgare*. Phytoecdysones are therefore likely to be ancient molecules that played—and still play—important roles in defending very resilient groups of plants. But caution is necessary in attributing defence roles to hormone mimics, and experimental proof for plant-derived ecdysones interfering with invertebrate development in nature is scarce. We may have to wait until a hormone-mimic pathway is engineered into a plant to test this.

Could mechanisms based on hormone mimicry affect reproduction in vertebrates, and, if so, could this act in nature as a defence against large herbivores? Many plant-derived compounds can interfere with various stages of reproduction in humans, perhaps the most famous being the now extinct *Silphium* plant, source of a powerful and historically important contraceptive (Riddle, 1997). But our willingness to attribute natural roles to contraceptives and abortifactants from plants may be a reflection of the catnip syndrome. Given the fast feeding processes of individual vertebrate herbivores, fast-acting defences are required. It may be primarily invertebrate herbivores with short lives which would be targeted by such hormone mimicry mechanisms in nature. That is, the effects of plant compounds on reproduction in vertebrates may be collateral. In the case of other terpenes the effects can be even worse than causing hormone imbalance. Some terpenes are mutagens and can cause organ failure or even cancer in mammals.

A radiomimetic norsesquiterpene from bracken fern

In between mono- and diterpenes in terms of size are the sesquiterpenes, and among these are the sesquiterpene lactones, compounds that are widespread in dicots but perhaps absent in monocots. Sesquiterpene lactones are generally sequestered as glycosides in plants and released upon wounding; in a parallel with some previous examples of phenolics, they provide further examples of reactive plant defence chemicals—as in the case of protoanemonin from buttercups (*Ranunculus* spp.; Ranunculaceae), they probably act as broad-spectrum toxins. This same compound is part of the formidable chemical defences of hellebores (*Helleborus* spp.), plants that survive the winter as evergreens on the forest floors of mountainous regions in Europe. The stinking hellebore (*H. foetidus*) in western Europe is one of these plants and, throughout the year, it has to resist the constant threat of chamois and deer. This plant, however, is generally not an aggresive weed. Although it uses potently reactive terpenes in its defence it is not as troublesome as the next and final example of leaf defence chemicals.

Bracken (*Pteridium aquilinum*) is one of the world's most widely distributed plants. Its rhizomes and the fact that it produces tough and leathery fronds each season makes this fern a serious agricultural weed. But among its hidden weapons are defence chemicals. Bracken produces variable levels of cyanogenic glycosides as well as the vitamin B1-destroying enzyme thiaminase. The fern also contains a remarkable terpene derivative called ptaquiloside. Like the levels of cyanogenic glycosides, the levels of this compound in bracken are very variable, ranging from undetectable to up to a mass of 1% of the dry weight of the fronds (Yamada et al., 2007). Farmers putting cattle out in the spring on the coastal hills of Wales are well aware of the dangers that the soft young fern fronds present. Eating this fern can reduce bone-marrow activity and cause retinal degeneration in both cows and sheep; when fed to

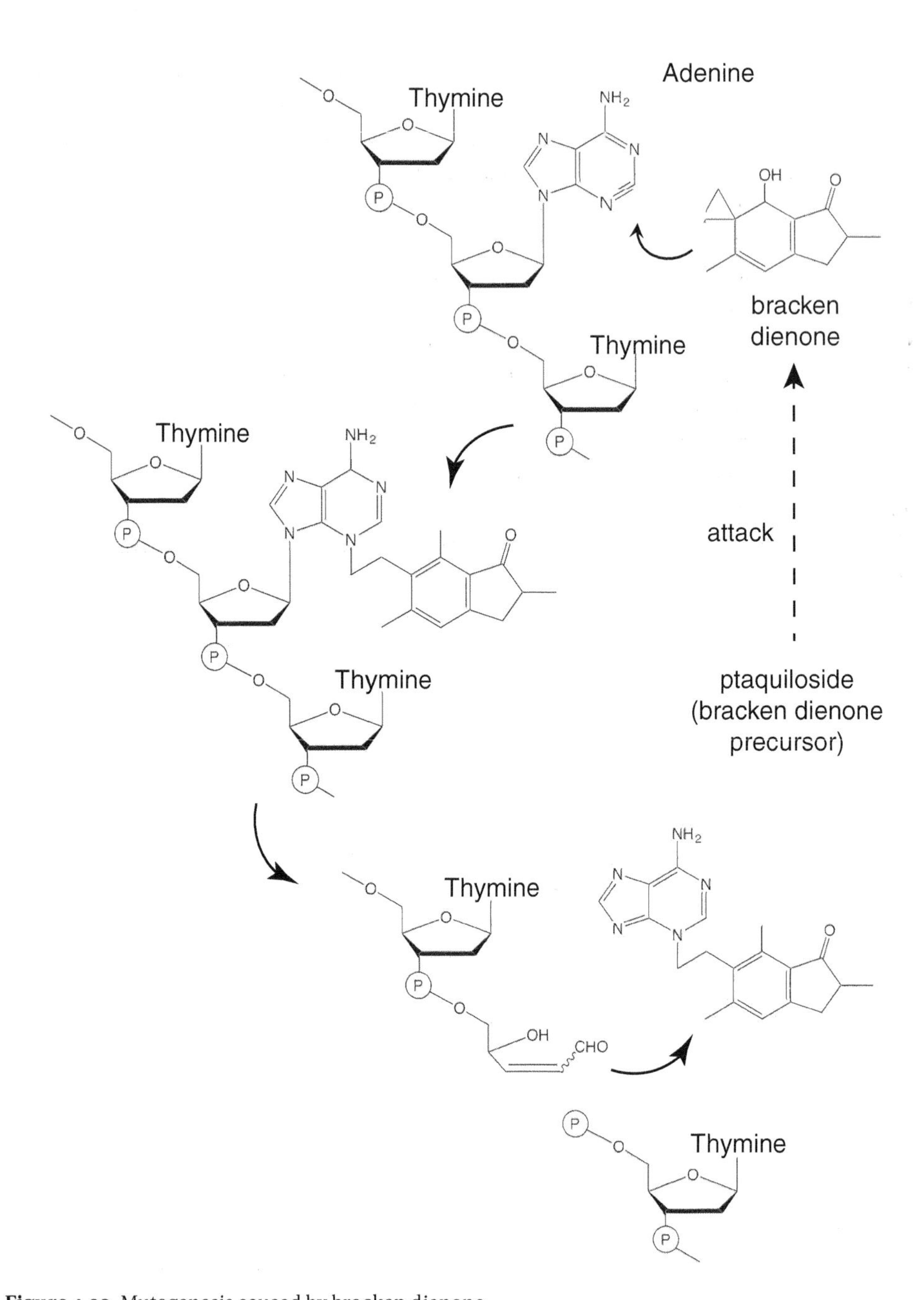

Figure 4.20 Mutagenesis caused by bracken dienone.
Damage to bracken fronds causes the deglucosylation of the precursor compound ptaquiloside to release bracken dienone, a molecule with a reactive cyclopropane ring which can react with a nitrogen atom in adenine. The adduct between the dienone and adenine then decays, causing the cleavage of the DNA backbone (after Yamada et al., 2007).

calves, bracken has been shown to cause tumours in the ileum. All these ailments have been ascribed to ptaquiloside, a norsesquiterpene (a sesquiterpene lacking one carbon) glucoside. The presence of ptaquiloside makes bracken a potentially carcinogenic plant. This is because ptaquiloside is a relatively unstable glucoside and readily loses its glucose to release bracken dienone, a powerful electrophile that can react with and cleave DNA. Interestingly, this cleaving activity is somewhat DNA sequence specific.

Bracken dienone reacts with a nitrogen atom in adenine, a reaction that destablizes DNA and leads to chain cleavage (Figure 4.20). One of the DNA sequences that was found to be most readily attacked was 5'-AAAT, where the underlined adenine (A) residue just 5' to a thymidine (T) is cleaved off (Yamada et al., 2007). Such nucleotide sequences are very common in the DNA of all organisms. But the fact that the dienone is released more readily from the glucoside under mildly basic conditions might suggest that it may be particularly effective against some insects with high pH guts, and it might also function against pathogens. Bracken is one of a select group, a plant containing a radiomimetic defence chemical (Evans, 1968). Aided by an invasive growth strategy and strong defences, this fern has conquered more than its fair share of the world.

5

Inducible defences and the jasmonate pathway

Nothing is static in a leaf that is under attack. Within seconds of being damaged, the signal pathways that will co-ordinate defence responses start to operate. In the minutes and hours that follow, the levels of chemicals present in the undamaged leaf begin to change and genes that are normally silent are activated. If attack is severe then defence responses will extend into distal tissues. Remarkably, some damage-activated genes encode proteins that have no function in the healthy leaf and only operate to the plant's advantage when they are in the herbivore's digestive system. A large proportion of these proteins reduce the herbivore's ability to procure amino acids and a further subset of leaf defence proteins block the incorporation of amino acids into animal proteins. Additionally, some plants target protein synthesis by producing non-protein amino acids. What are these defences and what controls their inducibility? Since inducible defence consumes resources, how is it connected to growth? When did the mechanisms that control inducible defences evolve?

Inducible proteins that deplete energy and essential nutrients

As the leaf forces the herbivore to expend energy to detoxify its small molecules, other molecular defences begin to work to interfere with the digestive and metabolic processes that the herbivore uses to procure nutrients and energy. Such defences can be particularly versatile weapons because they inactivate digestive proteins that are conserved deeply in both invertebrates and vertebrates. Indeed, at the cellular level, both groups of animals are very similar, sharing common requirements for vitamins and other nutrients. This provides excellent targets for plant-produced defence proteins. But this is all the more remarkable because mouths, stomachs, and intestines are harsh extracellular environments very unlike those in plant cells in which the proteins are born. Not surprisingly, many leaf defence proteins are very stable. Cooking is one of the few things that can inactivate the most resilient of them.

The literature on the ability of plants to destroy vitamins within the digestive tract of animals is not extensive, but there are examples among which is the production of thiaminases in horsetails (*Equisetum* spp.) and in bracken (Evans, 1976). Thiamin (vitamin B1) is an indispensable co-factor for pyruvate dehydrogenase (which catalyses the conversion of pyruvate to acetyl coenzyme A) and it is hard to imagine a more centrally important enzyme. But vitamins are not the most heavily targeted nutrients; this falls to the essential amino acids and leaves use several mechanisms to interfere with amino acid procurement, utilization, and metabolism. In some cases leaves produce non-standard amino acids that are either non-metabolizable or, through mimicking protein amino acids, become incorporated into proteins in the herbivore, thereby corrupting cellular function. Such defences are used, for example, in the leaves of *Arabidopsis* (Adio et al., 2011). Alternatively, many leaves produce inducible ribosome inactivating proteins (RIPs) that can block translation. Cereals such as barley do this (Reinbothe et al., 1994). But even more commonly, defence proteins work upstream of protein synthesis by blocking access to the leaf's amino acid content.

Essential amino acid destruction

The adoption of symbionts has saved folivores from some of the dangers of antinutritional defences. Aphids, for instance, depend on symbiotic bacteria to synthesize essential amino acids that might not, for a variety of reasons, be obtainable in sufficient quantity from the plant (Douglas, 1998; Hansen and Moran, 2011). But, for many leaf eaters, up to half of the amino acid types found in proteins either cannot be made at all by the animal or cannot be made in sufficient quantities. Usually listed as being arginine, histidine, isoleucine, leucine, lysine, methionine, phenylalanine, threonine, tryptophan, and valine, the majority of herbivores (both invertebrate and vertebrate) need them from their diets (Dadd, 1985; Reeds, 2000). A widespread defence strategy used by leaves is to target the animal's ability to procure one or more of these essential or semi-essential amino acids. Clearly, the plant itself cannot simply dispense with these molecules. By way of example, the two subunit types of the abundant carbon fixation protein ribulose bisphosphate carboxylase/oxygenase (Rubisco) in *Arabidopsis* both contain very similar proportions of essential amino acids (~44% of total amino acid residues). These are very abundant and nutritious leaf polypeptides. But if plants could somehow reduce the availability of even one of the essential amino acids in these proteins they could greatly impact a herbivore. This is exactly what many if not most plants do, either by reducing the hydrolysis of leaf proteins or by destroying leaf-derived essential amino acids in the animal's gut.

Two of the amino acids that are frequently targeted by plant proteins within the digestive system of caterpillars are threonine and arginine. If we consider threonine depletion, and take tomato leaves as the example, it is clear that the enzyme that does this, threonine deaminase 2 (TD2), has evolved to be activated by a proteinase within the digestive tract of certain insects (Chen et al., 2008; Gonzales-Vigil et al., 2011). The two threonine deaminase genes in tomato encode single polypeptide chains that associate to act as tetrameric enzymes. Each individual subunit in the enzymes folds into two principal domains, the catalytic domain responsible for conversion of threonine to α-ketobutyrate, and the regulatory domain that functions to reduce the catalytic activity of the enzyme while it is in the leaf (threonine deaminases are involved in the synthesis of isoleucine but this requires only moderate catalytic activity most of which is supplied by threonine deaminase 1, TD1). When caterpillars attack tomato plants they ingest TD2 along with TD1, which is less stable than its counterpart and is probably inactivated rapidly. TD2, the more robust of the two proteins, passes into the insect midgut where proteinases from the caterpillar cleave the polypeptide in half in a proteinase-sensitive linker amino acid sequence (Figure 5.1).

Now, liberated from its regulatory part, the TD2 catalytic domain resists further proteolytic attack and gets to work deaminating threonine—a process that is facilitated by the alkaline pH of the midgut at which the enzyme is at its pH optimum. So the amount of dietary threonine available to the insect is reduced and food that would otherwise be of good quality is now suboptimal. TD2 is not active against all insects and while the cleavage process that activates the enzyme occurs in lepidopterans it does not appear to be present in beetles which may lack the serine proteinases necessary to cleave the inter-domain linker sequence (Chen et al., 2008).

Enzyme inhibitors

The leaves of many species produce a battery of inhibitors of digestive enzymes to function alongside the enzymes that destroy micronutrients and essential amino acids. But what kinds of molecules would be the primary targets for such inhibitors? Levels of lipids such as triglycerides are typically very low in leaves. But the same is not true for starch and proteins. Here, a window of opportunity exists for the leaf to defend itself and, unfailingly, it is utilized. Starch-degrading

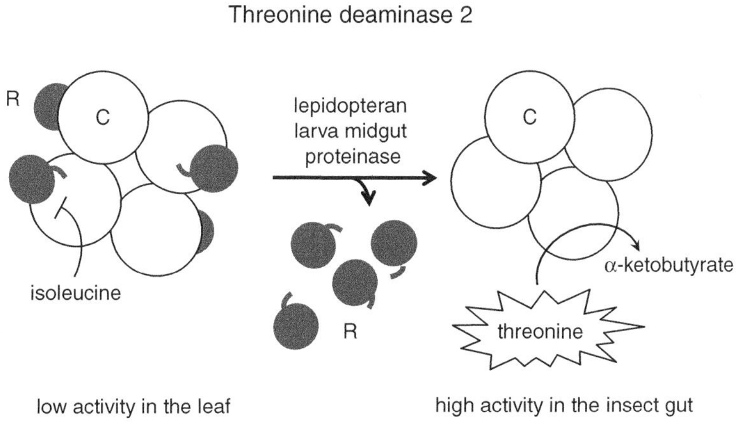

Figure 5.1 A threonine-destroying enzyme that is activated by insect herbivores. Threonine deaminase 2 (TD2) is a damage-inducible enzyme from tomato leaves that can deaminate the essential amino acid threonine, converting it to α-ketobutyrate. Each subunit of the TD2 enzyme has two domains encoded in a single polypeptide. The regulatory domain 'R' reduces the activity of the catalytic domain 'C' when the intact protein is stored in the leaf. When leaf contents containing TD2 are ingested by caterpillars, enzymes in the midgut cleave off the R domain so that the catalytic domain is fully active. The processed enzyme shown on the right has been crystallized and is known to be tetrameric, a tetrameric form in the leaf is probable. Adapted from Gonzales-Vigil et al. (2011).

amylases are targeted by amylase inhibitors while the activity of digestive proteinases is blocked by proteinase inhibitors. These latter proteins are, arguably, the prime examples of proteins made by plants to interfere with digestion in animals. Instead of destroying amino acids, these proteins work at another level—to block protein digestion so that amino acids cannot even be released from polypeptides. What we know about proteinase inhibitors is largely due to the work of C.A. Ryan who discovered that the feeding of Colorado beetles (*Leptinotarsa decemlineata*) on the leaves of tomato caused the accumulation of these proteins (Green and Ryan, 1972). Although it was already clear that plants could make proteinase inhibitors, the discovery was exciting because, among other things, it showed that the levels of these proteins increased strongly in leaves in response to the feeding of an insect. Furthermore, and suggesting roles in defence, the proteinase inhibitors in question were inactive against plant-derived enzymes and, instead, only inhibited digestive enzymes from animals. They do this remarkably well by forming extremely tight non-covalent interactions with the proteinase catalytic sites. Why these proteins have such broad roles in defence is easily seen.

Typsin and chymotrypsin, two of the digestive proteinases that were found to be targeted by inducible proteinase inhibitors from potato and tomato leaves, occur widely in both invertebrates and vertebrates. Potato and tomato leaves produce multiple forms of trypsin and chymotrypsin inhibitors with the best-studied being proteinase inhibitor I and proteinase inhibitor II. Both of these small, highly stable proteins are made when the leaf is damaged and are then stored in the vacuoles of leaf cells until the leaf is ingested by a herbivore. Once inside the animal, inhibitor I inhibits chymotrypsin whereas inhibitor II, an elegant invention of nature, is 'double headed' and can block both trypsin and chymotrypsin activity (Ryan, 1990; Jongsma and Beekwilder, 2008).

The production of proteinase inhibitors appears to be very widespread in plants, so most folivores have to live with them. To their peril, obtaining enough amino acids from the plants they feed on requires that they first expend some of their own valuable protein in the form of enzymes such as trypsin and chymotrypsin. However, if these animal proteinases are inhibited by the plant's proteinase inhibitors and then fail to cleave dietary proteins, the animal will begin to sense that it is not getting enough amino acids in its diet. The response, hardwired into the animal's physiology, is then to secrete yet more proteinases, proteins that inevitably contain essential amino acids. Again these are inhibited, further depleting the herbivore's reserves of amino acids. This, of course, takes time, so one might argue that such a defence might not be good against vertebrates—they might not associate a particular species of leaf with this type of defence. But no-one who has consumed uncooked plant material rich in proteinase inhibitors quickly forgets the surprisingly rapid discomfort that this causes. Furthermore, we know that proteinase inhibitors are part of the plant's defence arsenal because, being the products of single genes, their levels can be selectively modified through the use of molecular genetics. Indeed, experiments in several plant species have confirmed that overexpressing proteinase inhibitors in leaves leads to the reduced growth of insects (Johnson et al., 1989; Zavala et al., 2004). These types of experiments have also shown that the impact of producing proteinase inhibitors on plant growth can be low (Zavala et al., 2004; Hartl et al., 2010) so there is an obvious interest in exploiting these kinds of proteins in pest control (Schlüter et al., 2010).

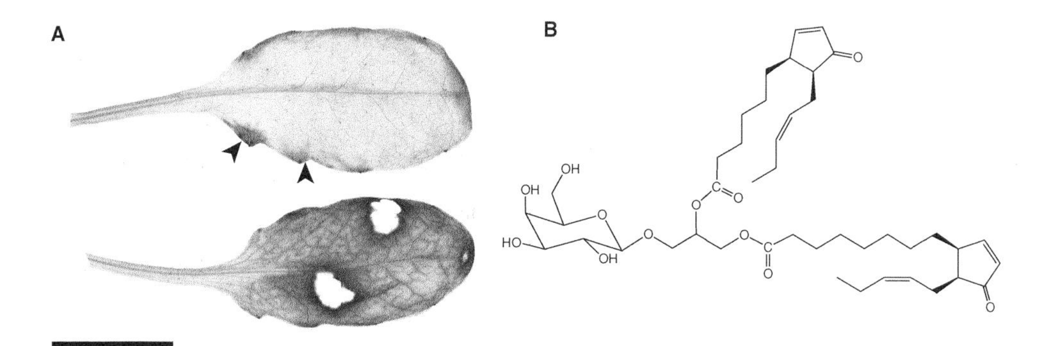

Figure 5.2 Highly inducible defence molecules in *Arabidopsis* leaves. Defence proteins and defence metabolites accumulate simultaneously when leaves are wounded by chewing herbivores. (A) Plants were engineered to express the promoter of the *VEGETATIVE STORAGE PROTEIN 2* (*VSP2*) gene coupled to a gene encoding a reporter enzyme (β-glucuronidase) that converts a colourless substrate to a blue precipitate. The upper leaf was from an undamaged plant whereas the lower leaf had been attacked by larvae of the Egyptian cotton leafworm (*Spodoptera littoralis*). The dark staining indicates activation of the VSP2 gene and arrowheads show basal promoter activity in an undamaged leaf. Scale bar = 1 cm. (B) An arabidopside. These lipid derivatives are highly inducible at the edges of wounds where their concentrations can increase to >1000-fold those in undamaged leaves. Unlike VSP2 which accumulates in the timescale of hours after wounding, arabidopsides accumulate in seconds to minutes. Both VSP2 and arabidopsides have established roles in defence (Liu et al., 2005; Glauser et al., 2009). Image: S. Stolz and E. Farmer. (See Plate 9.)

'Duality is the very essence of war . . . in war the power to use two fists is an inestimable asset' (Liddell Hart, 1944, p. 21). Complicating things still further for the herbivore is the fact that there may be synergy between enzyme inhibitors and small defence molecules. An example of the intervention of small molecules was seen in the leaves of tobacco where the effects of proteinase inhibitors were found to be accentuated by production of nicotine (Steppuhn and Baldwin, 2007). Concerning starch degradation, amylase inhibitors themselves can either be proteins or small molecules such as flavonoids (Franco et al., 2002; Lo Piparo et al., 2008). This pattern is very frequently the case as shown in Figure 5.2 for two highly inducible defence molecules in *Arabidopsis*, one a protein and one a fatty acid derivative. These are both highly wound inducible, although they may accumulate over different areas of a leaf that has been attacked by a herbivore. So, with each bite of a leaf, small molecules and proteins work in concert to target the process of digestion. Finally, making things even more difficult for the herbivore is the fact that many small defence molecules and many defence proteins are damage inducible. The more a leaf is attacked, the more it is defended.

The moving defence horizon

The healthy, unattacked leaf usually contains a resting level of phytoanticipins. If this chemical blend stayed stationary during attack it might be easier to evolve ways to overcome it. But plant defences are highly dynamic (Karban and Baldwin, 1997). Indeed, it seems to be critically important that the defence barrier changes during attack, and this probably helps to keep the advantage in the hands of the leaf. Changing the defence horizon is an effective way of complicating the attacker's life—an army equipped to fight in a hot desert might be unequipped to fight on snow. But there are further advantages in having inducible defences. For example, many defensive chemicals are used by specialized herbivorous arthropods as feeding or oviposition attractants (e.g. Renwick, 2002), so releasing them only when absolutely necessary would reduce visits from unwanted organisms. The same can be said for attack-inducible volatiles that alert predators to the presence of feeding herbivores; this would work only if false alarms due to constant production of such molecules were avoided. Then there is the fact that building most defences requires useful resources, this being especially so for protein-based defences that require much nitrogen to build. Yet another hypothesis, and one that is difficult to test, has been suggested over the years in several versions by a number of authors. This is based on the fact that, in their uninduced state, the leaves of some plants are less defended than fruits on the same plant. D.H. Janzen saw the vegetative tissues of some plants as bait to attract vertebrate herbivores to plants in order for them to disperse seeds (Janzen, 1986). As the bait is eaten it would become less attractive, thereby ultimately sparing the plant to complete its lifecycle.

Just as the levels of chemical defences and defence proteins increase upon attack, the production of specialized defence cells can be induced by herbivory. Examples of herbivore-inducible defences range from silification in grasses, to sting development in nettles, and to thorn production on acacias, to name but a few (McNaughton and Tarrants, 1983; Pullin and Gilbert, 1989; Milewski et al., 1991). Thus, induced responses to herbivory go well beyond simply augmenting the levels of a few chemicals. But how is the moving defence horizon controlled and what stops a leaf overproducing defences? We now know that the control of inducible chemical and physical defences is due in large part to the activity of a wound-inducible regulatory lipid called jasmonic acid. This plant-produced compound plays an underpinning role in regulating the majority of herbivore-inducible leaf defence responses. How do herbivores try to avoid activating the production of this molecule in leaves?

Activating inducible defence: the importance of having good teeth

Leaf tissues need to be crushed in order to strongly stimulate defence gene activation. If leaf tips or edges are severed with sharp scissors, or if the lamina is pricked with sharp pins, the resulting defence responses are not as strong as when cells (particularly those in the vasculature) are crushed. In keeping with this, countless invertebrates cut neat circular holes in leaves and this behaviour is clearly not fortuitous. First, it involves minimal movement on the part of the animal. But cutting circular holes in leaves has another advantage; the insect removes the maximum amount of tissue but leaves the smallest possible length of cut edge. Minimizing cut edges has been found to reduce the activation of defence in leaves (Reymond et al., 2000). Furthermore, most leaf-chewing insect larvae carefully avoid crushing cells while they feed; they can cut through a leaf at least as cleanly as can a human with a fresh razor blade. In fact, everything about the way most leptidopteran larvae feed reduces defence-related gene expression to a minimum. But this depends on having sharp mandibles. By maintaining their mouthparts in good condition, tissues can be sliced finely and the cut edges of leaves left with the minimal crushing of cells and the minimal release of plant-derived elicitors. Additionally, small pieces of food are easiest to digest. Many leaf defences come into play at this point. Abrasive substances like calcium oxalate crystals or silica enable the leaf to wear down the sharp facets of mandibles (Korth et al., 2006). These common defences can turn mouthparts into what the herbivore does not want: defence gene-activating devices.

Is a physical wound sufficient to provoke the full defence response?

How much of the leaf's response to being fed on by a chewing insect depends on damage to the leaf and how much depends on factors emanating from the insect, for example from its oral secretions? One way to address this is to compare responses to mechanical wounding and insect damage at the molecular level using hundreds or thousands of markers such as messenger RNAs. In some cases this approach has shown that mechanical wounds result in a gene expression programme rather similar to that triggered by an insect (Reymond et al., 2000). Additionally, experiments with a mechanical device that mimics insect feeding also triggered a response similar to that of insects (Bricchi et al., 2010). Indeed, this is consistent with the evolutionary perspective that having a defence system that is stimulated by any kind of mechanical damage keeps the plant in control. Whichever organism causes the damage also causes a defence response (Heil, 2009). However, in other cases, insects and physical damage are reported to have very different effects on wound responses in the plant since insect saliva contains elicitor substances (Alborn et al., 1997; Bonaventure et al., 2011). Two views emerge from the literature. One view favours elicitors from chewing insects as being major determinants of the outcome of a plant's response to damage. The other view is that it is exclusively the physical damage that chewing insects cause while feeding that activates the plant's response. In theory, mutants that have overactive defence responses could provide valuable information to help resolve the issue.

The ideal wound-mimic mutant would, in the absence of herbivores, display characteristics closely similar to that of an attacked plant—and I, for one, was initially sceptical that such a plant would exist. However, a mutant screen in our laboratory yielded an *Arabidopsis* plant that we named *fatty acid oxygenation upregulated 2* or simply '*fou2*' (Bonaventure et al., 2007a). When mRNA and protein levels in *fou2* were compared to those in wild-type plants, *fou2* was found to closely mimic a plant being fed upon by the cabbage butterfly (*Pieris rapae*) larvae (Bonaventure et al., 2007b). The gene expression pattern of the unwounded mutant resembled that of the insect-damaged wild type so closely that this prompted us to inspect plants for insect damage. There was none. So it is possible to at least recapitulate a large part of a wound response in the absence of a herbivore and its saliva.

As the field stands at present, it is likely that the relative contributions of physical damage and salivary components vary on a case-to-case basis, depending on the species of herbivore as well as on the plant. Generalizing, the smaller the herbivore the greater the impact of its oral secretions on cells in the host plant are likely to be. But physical damage inflicted by chewing herbivores surely plays a primary role in activating leaf defences. Our further analysis of *fou2* then revealed that the mutation resided in a gene encoding a vacuolar membrane (tonoplast) ion channel and the alteration in its sequence increased cation flux through the channel. This affected both calcium and potassium homeostasis in the leaf. A possibility to explain activation of leaf defences was that abnormally high calcium levels in the cytosol (and/or in organelles such as plastids) directly or indirectly stimulated the activity of genes that are regulated by the defence hormone jasmonate (Beyhl et al., 2009). Prior to looking at jasmonate in detail, it is necessary to outline the initial discovery of its regulatory properties.

The jasmonate pathway

Going back in time, the observation that proteinase inhibitors were inducible when an insect fed on tomato leaves (Green and Ryan, 1972) was remarkable because these proteins accumulated not just in the leaf that had been damaged, but also elsewhere in the plant, in undamaged leaves. The first thing that this suggested to me, having previously completed a PhD in a laboratory working on animal cells, was that there might be a kind of 'second messenger', a small regulatory molecule made by wounded leaves that would activate proteinase inhibitor gene expression. As soon as possible I joined Ryan's laboratory to try to search for such a molecule—a search that was facilitated by clues several in the literature and elsewhere. At the time, a favourite candidate for the 'proteinase inhibitor inducing factor' in tomato leaves was oligogalacturonic acid, a polysaccharide derived from pectin in plant cell walls (Ryan, 1974). But in whichever way these oligogalacturonides were applied to plant tissues they did not activate proteinase inhibitor synthesis very powerfully and so it seemed worthwhile looking for alternatives. Already, experiments by Mary-Kay Walker-Simmons in the Ryan laboratory had shown that α-linolenic acid applied to leaves stimulated the synthesis of small amounts of proteinase inhibitors. Interestingly, however, the closely related compound linoleic acid was inactive, although the significance of this was unclear and the experiments remained unpublished. Nevertheless, reading the literature on plant metabolites that could be derived from linolenic acid (and that could not be derived from linoleic acid) yielded a publication on several cyclic fatty acid derivatives (Vick and Zimmerman, 1984). The smallest of the molecules mentioned, jasmonic acid (JA; Figure 5.3) obviously merited investigation.

At the time, jasmonic acid had been shown to have a number of biological activities in either promoting plant senescence (Ueda and Kato, 1980) or in inducing protein synthesis in pieces of plant tissue (Sembdner and Parthier, 1993). We needed to procure this substance and were able to obtain its methyl ester (methyl jasmonate) from a generous industrial supplier to the perfume industry. Upon opening the bottle the odour was instantly recognizable—as its name suggests, methyl jasmonate has the scent of jasmine flowers. This property of methyl jasmonate would later turn out to be interesting, although this was not foreseen when the first experiment was conducted. To start the experiment, a dilute suspension

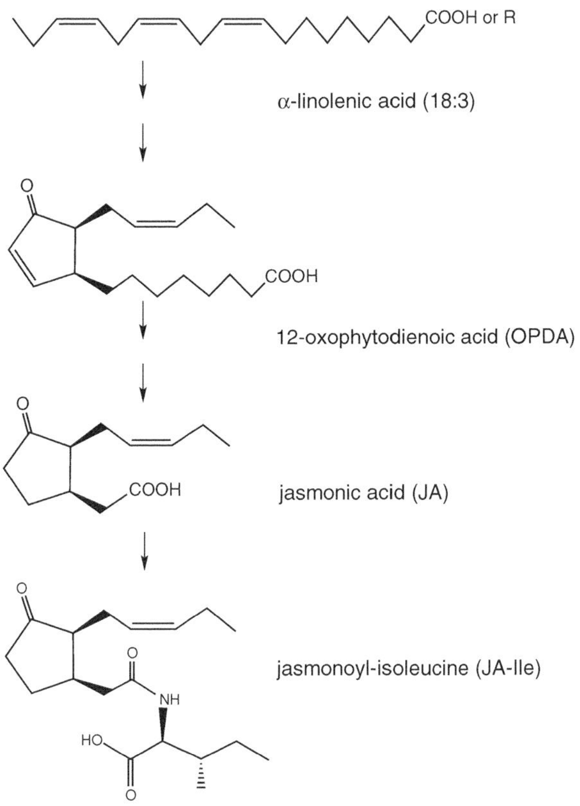

COOH or R

α-linolenic acid (18:3)

COOH

12-oxophytodienoic acid (OPDA)

COOH

jasmonic acid (JA)

NH

HO

jasmonoyl-isoleucine (JA-Ile)

Figure 5.3 Jasmonates and their precursors. Jasmonates originate from α-linolenic acid (often simply referred to as '18:3', i.e. with 18 carbons and three double bonds). 18:3 is converted to the jasmonate precursor OPDA, a molecule found in representatives from all land plant groups examined. By contrast, jasmonic acid (JA) and its derivatives such as jasmonoyl-isoleucine (JA-Ile) appear to be more modern and have yet to be identified rigorously from algae, bryophytes, and lycophytes. JA-Ile is a ligand for jasmonate responses in angiosperms and possibly other plant lineages. For additional details see Wasternack and Hause (2013).

of methyl jasmonate in a carrier solution was prepared. The control solution had already been tested and was known not to activate proteinase inhibitor production in the tomato leaves. Two days after initiating the experiment it was time to look at the results—they were so spectacular that they could have been missed rather easily. The production of proteinase inhibitors in the methyl jasmonate-treated plants effectively went 'off-scale'. Strangely, however, control plants in the same environment had also responded to the volatile and also made some proteinase inhibitors. This latter result could be explained by the fact that some of the methyl jasmonate sprayed on to the first series of plants had volatilized and had acted as an airborne signal to activate proteinase inhibitor production in the nearby leaves of the control plants. Indeed, tiny drops of dilute methyl jasmonate solution in ethanol could be placed on a cotton swab and, in closed containers, this powerfully elicited proteinase inhibitor production in tomato leaves. The effect was remarkably potent. Half-maximal inhibitor production was seen at around two nanolitres of methyl jasmonate per litre air volume (Farmer and Ryan, 1990). Our interpretation of the results was that the tomato leaves took up volatile methyl jasmonate through the stomata and that, once in the leaf, the molecule was likely to be demethylated to release jasmonic acid. Moreover, the potency of induction of the proteinase inhibitor proteins was such that we envisaged a mechanism that would transport jasmonate efficiently over small distances between cells within the leaf (Farmer and Ryan, 1992). Jasmonate was almost certainly a component of the wound signal in tomato leaves and this triggered additional research.

Not long after the initial realization that defence protein accumulation could be elicited by treatment with methyl jasmonate, the group of Meinhard Zenk showed that the compound could induce the synthesis of defence-related chemicals in 36 different cell suspension cultures (Gundlach et al., 1992). Quickly, a wide variety of other defence-related proteins and chemicals were found to be jasmonate inducible and many of these were also known to be wound inducible, strengthening the link between wounding leaves and activating inducible defences.

Today, the list of defence-related features that can be induced by treating plants with jasmonate is long. Jasmonates increase the levels of a wide variety of chemicals and proteins in plants, ranging from paclitaxel (an antitumour diterpene alkaloid) in the cells of Pacific yew (Yukimune et al., 1996) to glucosinolates in the Brassicaceae (Textor and Gershenzon, 2009; Brader et al., 2001), threonine deaminase in tomato leaves (Chen et al., 2008), to VEGETATIVE STORAGE PROTEIN 2 (VSP2, for which the activity of the wound-inducible promoter is shown in Figure 5.2A) in *Arabidopsis* (Liu et al., 2005). Other examples of jasmonate-stimulated phenomena include gum production (gummosis) in peach (Saniewski et al., 1998), laticifer differentiation in rubber plants (Hao and Wu, 2000), and the stimulation resin-based defences in gymnosperms such as a spruce (Erbilgin et al., 2006; Bohlmann, 2008). Furthermore, whole defence-related structures such as trichomes (Traw and Bergelson, 2003; Kobayashi et al., 2010), and even extrafloral nectaries that attract ants to plants, could be induced by jasmonate treatment (Heil et al., 2001). Finally—and opening up potential uses in agriculture—jasmonate could be applied to plants in the field to elicit defences (Ryan and Farmer, 1999; Thaler, 1999).

The biological roles of jasmonate

The fact that methyl jasmonate or jasmonic acid applications to leaves stimulated so many inducible defence-related phenomena begged the question of what would happen if plants could not produce or respond to this compound. Mutants in both jasmonate biosynthesis and signalling were needed and these soon started to emerge from a number of laboratories. The first of these genetics experiments was conducted with tomato plants and they revealed the increased susceptibility of jasmonate mutants to tobacco hornworm larvae (*Manduca sexta*) (Howe et al., 1996). Additionally, *Arabidopsis* plants lacking the triunsaturated fatty acid precursors of jasmonic acid were greatly more susceptible than wild-type plants to a fungus gnat (*Bradysia impatiens*, Diptera) (McConn et al., 1997). Since then, numerous other experiments have shown that a lack of ability to make jasmonate makes plants extremely susceptile to many types of invertebrate herbivore (Howe and Jander, 2008).

As shown in Figure 5.4, when generalist chewing insects are placed on plants that cannot make jasmonates they gain weight far more rapidly than those placed on wild-type plants. Invertebrates that are allowed to feed on the wild-type *Arabidopsis* rosettes show stereotyped feeding behaviours (which differ between species), but almost all have one thing in common: they concentrate most of their feeding on expanded leaves and do not succeed in feeding on younger leaves in the centre of the plant. As is typical, these young leaves are highly defended. They have high trichome densities, they produce high basal leves of certain defence proteins (such as VSP2) and they are rich in glucosinolates. However, when the ability of the plant to make jasmonate is specifically impaired, the insects instead concentrate

Figure 5.4 Insect growth on plants lacking the ability to make jasmonates. The plant on the left is wild type whereas the plant on the right is a jasmonate biosynthesis mutant (*lox2-1 lox3B lox4A lox6A*; Chauvin et al., 2013). Neonate *Spodoptera littoralis* larvae were allowed to feed on the plants for 11 days, at which point the mass of the insects that fed on the jasmonate mutant was greater than that of the insects that had fed on the wild type. Note the damage to the central growth region of the jasmonate mutant. Despite removal of the apical meristem these plants will achieve some seed production through the activation of lateral floral meristems. Scale bar = 1 cm. (See Plate 10.)

most of their attack on the youngest leaves of jasmonate mutants. They destroy these small leaves voraciously and, still concentrating on the centre of the plant, they even remove the normally highly protected apical meristem. Critically, energy distribution between the plant and the insect is changed. Whereas the wild-type plant retains most of its mass, jasmonate mutant plants are reduced to skeletons and disappear altogether if the insects are not removed. In parallel, insects on the jasmonate mutant gain mass far more quickly than those on the wild type.

Simple experiments like those with insects demonstrated the role of the jasmonate pathway in controlling resource availability to invertebrate herbivores. These experiments have been extended to vertebrates and initial results showed that plants which fail to produce or perceive jasmonate are preferred over the wild type by tortoises (Mafli et al., 2012). Experiments on agriculturally relevant herbivores are now awaited.

Jasmonate signalling in the attacked leaf

A few years after the discovery that jasmonic acid could stimulate defence responses in plants it was found that it is not free jasmonate that does this but modified forms in which jasmonic acid is conjugated to amino acids. Jasmonoyl-isoleucine (JA-Ile; Figure 5.3) was found to be a biologically active signal (Staswick and Tiryaki, 2004), as had been proposed previously (Krumm et al., 1995). When *Arabidopsis* leaves are wounded they start to synthesize jasmonic acid, the immediate precursor of JA-Ile, and this occurs remarkably rapidly. Jasmonic acid accumulation in a wounded *Arabidopsis* leaf is exponential, the first increases being observed after only 30 s within the site of wounding and ~15 s later in undamaged leaf tissue near a wound (Glauser et al., 2009; Chauvin et al., 2013). Shortly afterwards, jasmonoyl–amino acid conjugates such as JA-Ile accumulate and begin to do their job to stimulate defences (Koo et al., 2009; Chauvin et al., 2013). JA-Ile (and perhaps other conjugated forms of jasmonate) acts as a signal to alter the expression of genes in attacked plants, but how many genes are affected and what proteins do they encode? To address this question additional mutants in the pathway leading to the synthesis and perception of JA-Ile were needed and among the most useful of these are mutants in a plant gene called *corontatine insensitive 1*, or '*coi1*' in shorthand (Xie et al., 1998). This gene was named from a bacterial virulence factor called coronatine, a molecule that can act as a molecular mimic of JA-Ile to powerfully activate the jasmonate signalling pathway. The initial *coi1* mutant, *coi1-1*, was found in *Arabidopsis* and plants carrying this homozygous mutation do not respond to treatments with either coronatine, jasmonic acid or JA-Ile. We now know that this is because the protein that is made by the wild-type *COI1* gene is part of the core of the jasmonate perception process.

Once it had been established that the *coi1-1* mutant interfered specifically with the jasmonate pathway it could be used in a variety of experiments to study the effects of jasmonates without having to apply these molecules to plants. One question investigated was how many genes responded through jasmonate perception when butterfly larvae were allowed to feed on leaves? The larvae of the cabbage white butterfly *Pieris rapae* were left to feed on the leaves of wild-type and *coi1-1 Arabidopsis* plants for 3–5 h prior to analysing gene expression. Through this approach it was seen that caterpillar feeding affected the levels of mRNAs for >1000 of the ~27,000 genes in the *Arabidopsis* genome, and, of these, a surprisingly large proportion (range of 67–85%) were found to be regulated by the jasmonate pathway—nearly all of these being upregulated (Reymond et al., 2004). Less surprisingly, many of the genes that were activated via jasmonate signalling after caterpillar feeding encoded defence-related proteins such as VSP2, as well as enzymes involved in the synthesis of glucosinolates. In conclusion, jasmonate signalling was found to be required for the correct expression of more than three-quarters of the genes that were upregulated by feeding caterpillars. A smaller proportion of genes were regulated in other ways, for example in response to water loss from the insect damage sites.

How jasmonate works to unleash defences in leaves

There is now an emerging scheme for how jasmonate signalling works at the molecular level. Conjugates such as JA-Ile, made when a herbivore damages a plant, act in minute quantities to initiate a cascade of events leading to changes in the activity of genes. An early and critical step in this process is the binding of jasmonate conjugates directly to the COI1 protein (the protein for which the gene is mutated in the *coi1-1* mutant). The current picture is that COI1 bound to JA-Ile attracts a second type of protein called a JAZ protein. JAZ proteins are at the heart of jasmonate signalling and they appear to play most of their diverse roles while binding to and inhibiting the activity of other proteins. In the cells of healthy leaves these proteins are bound to certain transcription factors, proteins that help control the activity of genes (Figure 5.5). What COI1 bound to JA-Ile does is to pull dimeric JAZ proteins that are probably still bound to transcription factors on to its surface, so that the JAZ and COI1 proteins form a sandwich where JA-Ile is the filling.

Now that JAZ and COI1 proteins have been brought together, other protein partners that are associated with COI1 modify the JAZ proteins by ubiquitination. The addition of ubiquitin (itself a polypeptide) to other proteins usually condemns them to be destroyed and this is the case with the tagged JAZ proteins which are degraded rapidly and therefore cannot rejoin the transcription factors to which they normally bind. So, one of the first things that chewing herbivores do when they feed on a leaf is to cause a transient reduction in the levels of JAZ proteins. At this point, with JAZ protein concentrations in the cell lowered, the concentrations of free transcription factors are now increased and these freed transcription factors get to work immediately, accelerating the transcription of defence genes (Chini et al., 2007; Thines et al., 2007; Sheard et al., 2010).

Among the RNAs that accumulate most rapidly in response to wounding are those that encode JAZ proteins, one of which is JAZ10 (Yan et al., 2007). Increased levels of its RNA first detectable within about 5 min of wounding, so, in a herbivore-damaged leaf, its accumulation takes about ten times as long as does the accumulation of jasmonic acid and roughly four times as long as it takes to increase the levels of JA-Ile. In less than 1 h, defence proteins begin to accumulate. An insect therefore only has a limited time to feed on a leaf before its protein contents begin to change. Likewise, if a larger animal bites the end of a leaf and later returns to finish off the rest of the leaf it will surely encounter newly synthesized defence

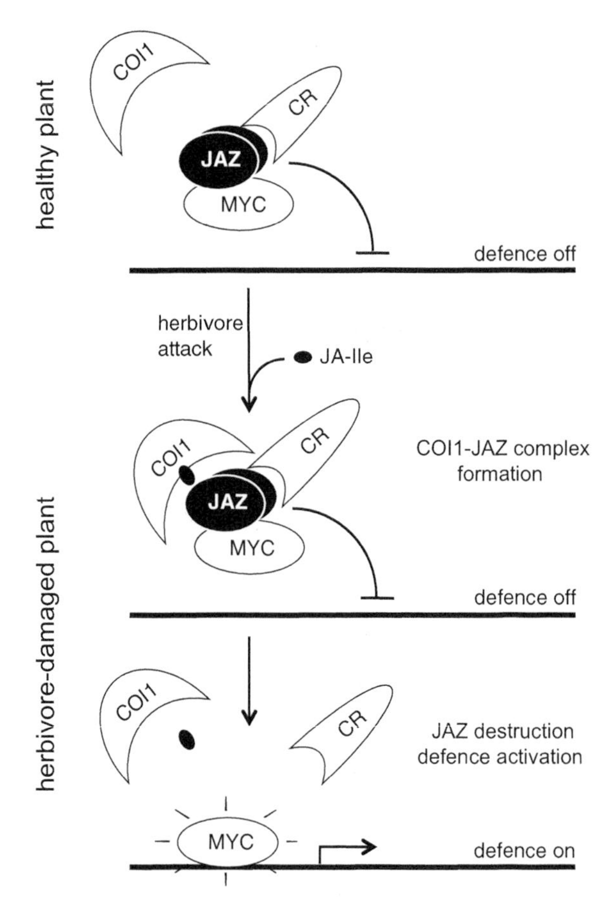

Figure 5.5 The core jasmonate signalling machinery. Jasmonate signalling is based on the interaction of two complexes: the COI1 complex, that can target proteins for rapid destruction, and the JAZ/MYC complex. In a healthy plant, JAZ proteins in combination with co-repressors (CR) bind to and inhibit the activity of transcription factors including those in the MYC family. Feeding herbivores stimulate the synthesis of active jasmonates such as jasmonoyl-isoleucine (JA-Ile, indicated as a black dot). This acts as 'molecular glue' to bring the two complexes together, leading to the destruction of JAZ proteins. Once these have been destroyed, MYC transcription factors transcribe genes necessary for the defence response, genes such as those encoding inducible molecules that interfere with digestion in the herbivore. These early events, represented here in a simplified form, take place in the nucleus.

compounds. However, once activated, the jasmonate pathway must somehow be kept in check since its overactivation would exhaust the entire plant. Now, two further factors come into account when the plant has responded to a wound. First, there are interconnected positive and negative regulatory loops that constrain the activity of the pathway (Figure 5.6). A second crucial factor is that, unless the plant is a small seedling, the effects of herbivore damage spread to only a limited number of leaves.

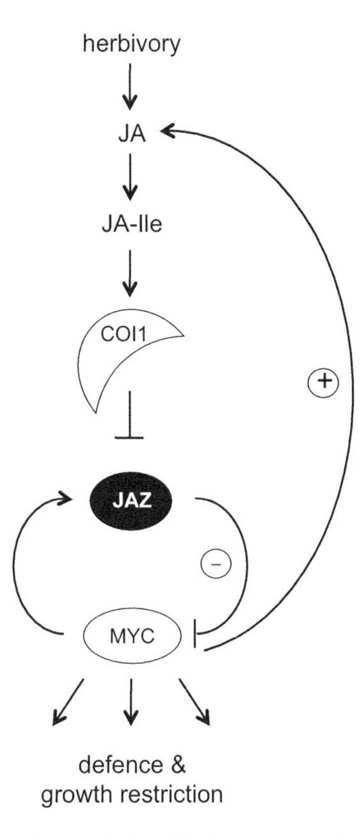

Figure 5.6 Regulatory loops control the activity of the jasmonate pathway in leaves. In *Arabidopsis*, jasmonates stimulate their own synthesis through increased jasmonate signalling as indicated in the regulatory loop marked '+'. A negative feedback loop, marked '–', is therefore needed to stop excessive activation of the jasmonate pathway. This is based on the fact that the MYC transcription factors enhance the production of JAZ proteins that then bind to MYC to inhibit jasmonate signalling. After Farmer (2007).

Signal propagation to distal tissues

What is initiated in the seconds after a leaf is bitten leads to changes that will affect the energy and time the herbivore needs to expend on eating—unless it moves off to a new leaf. As can be seen in any garden, caterpillars, slugs, and all manner of other invertebrates frequently remove a small amount of tissue, then move off to feed elsewhere, often on an entirely new, undamaged leaf. There are at least two likely reasons for this. First, small herbivores increase their chances of survival by separating themselves from visual and chemical evidence of their presence. Second, once a herbivore begins feeding it stimulates the synthesis of jasmonate and this activates defences near the wound. If the bites that the herbivore inflicts are sufficiently damaging, or if there are multiple herbivores feeding on the same leaf, then jasmonate synthesis and signalling are stimulated not only in the attacked leaf but also in distal parts of the plant. If this happens the herbivore will have limited the availability of good feeding grounds unless it moves to an entirely new plant.

To look at this in more detail the process of what happens when a leaf is bitten can be broken down to three principal steps. In the 'initiation phase', wound damage causes the production of moving wound signals. This is followed by the signal 'propagation phase', and finally there is a 'decoding phase' whereby the long distance signals arrive at cells and initiate the process of jasmonate production. Almost nothing is known about the initiation phase—what happens during cell damage to initiate the signalling process. At least in the case of a mechanical wound the strong activation of the jasmonate pathway probably depends to some extent on endogenous factors from crushed plant cells. Indeed, many studies support the idea that plants contain damage-associated molecular patterns ('DAMPS') that have evolved to initiate the wound signalling process, and increasing evidence suggests that among these are defence-activating peptides (McGurl et al., 1992; Huffaker et al., 2013). It is thought that these act early in antiherbivore defence signalling upstream of jasmonate synthesis (Farmer and Ryan, 1992). But we currently know so little about the initiation phase that we cannot be certain whether it is chemical signals generated upon wounding, or the physical events related to the wound, or, as seems most likely, both, that get the ball rolling.

Unlike for the still mysterious initiation phase, something is known about the propagation of long distance wound signals. Much of this comes from work on two species: tomato and *Arabidopsis*. These plants differ greatly; unlike tomato, *Arabidopsis* grows in a squat rosette form prior to flowering and plants that have been grown for 5 weeks have ~20 expanded leaves. They are ideal for studying leaf-to-leaf signalling in response to wounding. Furthermore, the pattern of vascular connections between leaves has been determined, and this is important in experimental design. To know which leaves are interconnected to other leaves through the vasculature, the direction of the spiral growth pattern in each individual plant must be determined. Fifty per cent of plants produce rosettes which spiral to the right, all others spiralling to the left. One can quickly count upwards from the oldest non-cotyledon leaf, leaf 1, so that each true leaf has its own identity (Figure 5.7). Wounding experiments can then begin.

When a single *Arabidopsis* leaf in the middle of the rosette is wounded by crushing a large area of the leaf tip, several other leaves show induced defence responses. Provided that one works with rosettes of the same age, these leaves always respond in the same way between plants. There is, of course, no visible response in the non-wounded leaves and, instead, one can measure the accumulation of jasmonic acid, JA-Ile or jasmonate-responsive transcripts. Therefore in a typical experiment leaf 8 is wounded and, 1 h later, each expanded leaf on the plant is harvested and analysed for its content of a highly wound-inducible mRNA known to be controlled by jasmonate signalling (for instance transcripts from the *JAZ10* gene). In this type of experiment, four leaves that respond particularly well are leaves 5, 11, 13, and 16 (Figure 5.7). If, however, leaf 9 is wounded, then good defence responses are seen in leaves 6, 12, 14, and 17. Underlying this predictable response pattern are vascular connections between the damaged leaf and the undamaged leaves. The important thing here seems to be extent of leaf damage. If a number of small herbivores are feeding simultaneously on a single leaf, or if

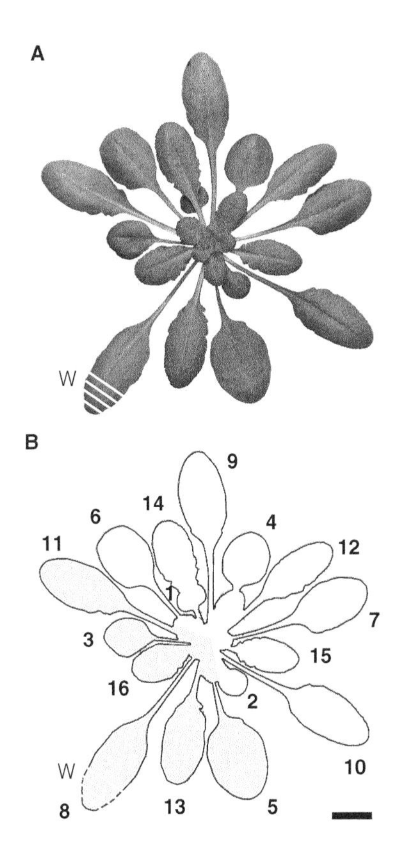

Figure 5.7 Wound-response domains in a small plant. (A) The rosette of a 5-week-old *Arabidopsis* plant. The 'W' indicates a severe wound on leaf 8. In response to the wound, defence responses under the control of the jasmonate pathway are induced rapidly in several leaves. (B) Outline of the rosette with this wound response represented by activity of the *JAZ10* jasmonate response gene at 1 h after wounding. The domain of *JAZ10* expression is highlighted in grey. Scale bar = 1 cm. Adapted from Mousavi et al. (2013). Image: S. Mousavi and E. Farmer.

many cells are crushed by a vertebrate, the signals produced in response to wounding travel to the centre of the plant, then 'overspill' into certain other leaves. However, a single insect feeding stealthily will set off the production of wound signals but they may not get further than the centre of the rosette. Nevertheless, most importantly for the plant is that signals reach the region of the apical meristem from which new foliage is produced. This induces further defence in this region as well as having long-term consequences for plant growth.

The nature of long wound distance signals

Experiments to estimate how quickly long-distance wound signals travel though a plant were only initiated after the discovery that the levels of jasmonic acid rose much faster than expected after wounding in both the wounded leaf itself and in distal leaves. First attempts placed the signal velocity in the range of a few centimetres per minute (Glauser et al., 2008). This then provided a working timeframe upon which further experiments could be based, and subsequent measurements of JA-Ile accumulation also placed long-distance signal speed estimates in the range of a few centimetres per minute (Koo et al., 2009). Further leaf-to-leaf information transfer measurements, this time by taking into account interleaf vascular connections, yielded speed estimates of 3–8 cm/min (Glauser et al., 2009; Chauvin et al., 2013). The wound signal, then, moves from leaf to leaf at a speed roughly equivalent to the walking speed of a small invertebrate herbivore.

Next is the question of the nature of the long distance signal that activates the jasmonate pathway. This has been controversial for years but what is now clear is that at least part of the long distance signal is electrical. This stems in part from initial work on the induction of proteinase inhibitor genes in wounded tomato leaves where proteinase inhibitor production was found to correlate with the generation of electrical activity (Wildon et al., 1992). The question that remained, however, was whether there was a causal link between electrical signalling and defence activation. A genetic approach was adopted using *Arabidopsis* and this has shown that the signals activating the jasmonate pathway at a distance from the wound are indeed electrical. Supporting this, the apparent velocity of the electrical activity measured was close to the speed estimated by measuring the accumulation of jasmonate: also 3–8 cm/min. Remarkably, the production of these long distance signals depends on genes related to those that work in fast synapses in vertebrates (Mousavi et al., 2013). At present it seems likely that long distance signal initiation and/or propagation depends on plasma membrane depolarizations that move over long distances through cell files in the vasculature. However, the full wound signal may well be more complicated. Depolarizations of the plasma membrane are quite likely to be associated with other types of signal such as cellular pressure changes, and it is almost certain that jasmonate or a derivative moves over at least short distances between cells. More work is needed to bring this area to a respectable level of knowledge.

The phloem is proposed to be a major route of wound signal propagation, although xylem tissues are also candidates (reviewed in Fromm and Lautner, 2007). Interestingly, the first jasmonate produced in undamaged leaves on a wounded *Arabidopsis* plant appears to originate from developing xylem cells (Chauvin et al., 2013). Perhaps these cells lie along or near a path of long distance wound signal propagation. The last step in response to the arrival of a long-distance wound signal is the decoding step necessary to initiate jasmonate synthesis in plastids. The plastid, where jasmonate synthesis begins, is thus a good bet to be among the organelles that first respond to the arrival of the incoming wound signal initiated by a feeding folivore. Being furthest from the events that initiate wounding, the decoding phase may turn out to be the most experimentally tractable phase: this research area is expected to develop quickly.

Jasmonate and growth

Critically, most long distance signals made in response to feeding herbivores travel through the leaf to regions where they can affect future growth—regions of cell proliferation.

Jasmonate signalling affects growth in these cell populations as much as it activates defences in and around them. Growth and defence are interlinked, and for any organism, fast growth means fewer resources for defence. This can be manifest at various scales: from populations down to single cells. Looking first at the overall growth strategy of plants (and, for the present, putting wound responses aside) it can be seen that each species is faced with a growth/ defence 'dilemma' (Herms and Mattson, 1992). How much of its resources it devotes to its basal defence rather than to growth helps to determine how much it is attacked. This has been captured elegantly in the 'resource availability hypothesis' by Coley et al. (1985). The idea is that plants which are adapted to grow on nutrient-rich soils invest relatively little in defence. These plants have fast growth rates and short life cycles, and they do not spend a long time in the environment. By contrast, plants adapted to growing on nutrient-poor soils grow slowly, are long-lived, and they invest more heavily in defence. These latter plants tend to build up high levels of chemical defences and many have highly fibrous leaves. The bulk of fast-growing plants comply well with the model, investing relatively little in defence and often suffering high rates of herbivory. But there are some exceptions, among these being plants that have evolved close associations with vertebrate herbivores from which they get most of their nutrients, allowing them both to grow quickly and to make nitrogen-based defences. It is a trade-off between growth and defence for all but the richest.

Beyond the level of a plant's overall growth strategy there is also the effect of the environment on its development. For example, factors such as light quality and quantity impinge on growth and also affect defence. This is seen in the saplings of many forest trees that grow extremely slowly in the shade, biding their time and defending themselves well through the sapling stage until they eventually obtain enough light to allow them to grow faster and escape the reach of ground-based herbivores (Coley et al., 1985). Competition between plants, for example in dense forests, can cause aetiolation and shade-avoidance such that particular regions of the shaded plant show accelerated growth. Again, this comes at the expense of defence and it is known that this is due in part to reduction in the activity of the jasmonate pathway in extendable regions such as petioles (Ballaré, 2009; Cipollini, 2005; Moreno et al., 2009). The more the petiole extends the leaf blade in the search for light, the less it is likely to be defended during this process. Selective petiole feeders (several monkey species, for example) might, at least in theory, prefer to eat the petioles of shaded leaves.

Finally, and independent of other environmental factors, are the short-term effects on growth that are activated by wounding. This is seen throughout the lives of plants, starting at the seedling stage where even small wounds both excite defence responses and inhibit growth. What is interesting about wound-induced growth responses is that they are not simply growth inhibition. Here things are more complicated than a growth/defence dilemma and instead it might be better to envisage a growth/defence union. The new leaves emerging from the previously damaged plants may differ more or less subtly to their predecessors on the previously unwounded plant (e.g. Björkman et al., 2008). First, these leaves can have higher trichome densities than the equivalent leaves of unwounded plants. Second, severe wounding can also affect the size of subsequently produced leaves which tend to be smaller than leaves from unwounded plants and may be borne on noticeably shorter petioles, making the plant more compact. This short-term growth remodelling increases the level of structural defence in the plant and, in parallel, levels of defence chemicals and proteins in the new leaves are also increased. The overall restriction of growth is an important means of allocating resources to defence but it goes beyond this. As mentioned in the introduction, a plant that has suffered attack will often carry (sometimes to its advantage) the effects of this later in life.

There is evidence that the slowing of growth after wounding involves the tight interplay of the jasmonate signalling machinery with growth-related transcription factors and their associated repressor proteins. Part of what we know about this is shown in Figure 5.8. One sees that the regulation of wound-induced jasmonate responses and growth processes are highly interconnected and this has many implications. For instance, the negative association between growth and jasmonate-controlled defence is presumably exploited by herbivorous arthropods such as gall-forming insects. The mechanisms they use to stimulate cell division in the interior of galls provide them with their food: rapidly growing, undifferentiated cells. It is likely that by forcing fast growth, gall-formers also reduce the defence capacity of the cells they eat.

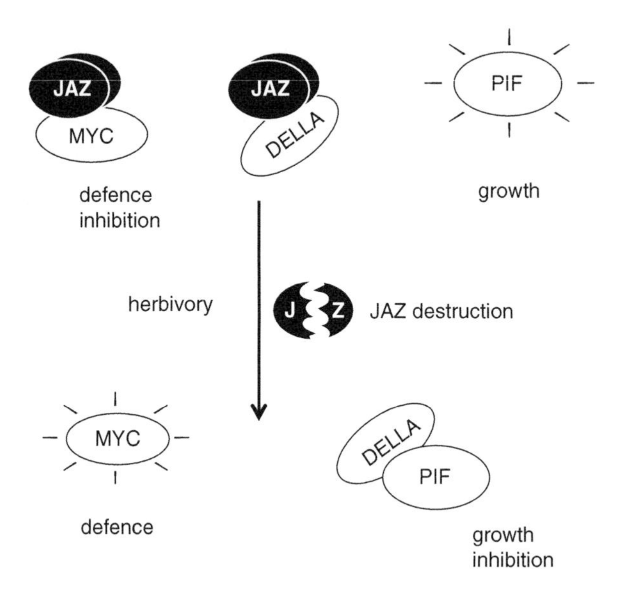

Figure 5.8 A proposed mechanism to link wound responses and growth. JAZ proteins bind to both MYC transcription factors and to small growth-regulating polypeptides called DELLA proteins. When herbivores attack leaves JAZ proteins are destroyed, liberating MYC transcription factors to transcribe defence genes. At the same time DELLA proteins are also liberated, but they now bind to other growth-promoting transcription factors called PIFs. Growth slows as soon as DELLAs sequester PIFs. The exact role of all proteins in this first-generation model is still under investigation. Adapted from Hou et al. (2010) and Yang et al. (2012).

The evolution of jasmonic acid-based signalling

The appearance of the jasmonate signal pathway was a major event in the evolution of plants as it gave them new ways of regulating inducible defences and of linking defence and growth. So it is therefore of interest to know when the ability to use jasmonates as defence signals first arose. Evidence is now emerging but the picture is not yet clear. Complicating the matter, some of the key jasmonate signalling proteins in angiosperms are derived from more ancient molecules found in early plant groups that appear to lack jasmonate signalling. So when did jasmonates themselves first make their appearance?

OPDA (12-oxo-phytodienoic acid), a central precursor in the synthesis of jasmonic acid (Figure 5.2), itself has a long history in land plants and has been found in the bryophytes. Indeed, this molecule is detectable in liverworts, in lycophytes, and in ferns, as well as in a variety of gymnosperms. But the presence of OPDA only points to the fact that a plant has the first part of the enzymatic machinery necessary to make a jasmonic acid precursor. This does not mean that a plant has jasmonate signalling so one has to search for jasmonic acid itself. But searches for jasmonic acid and jasmonoyl–amino acid conjugates are usually more difficult than those for OPDA since the concentrations of these molecules can be very low (*Arabidopsis* is very atypical in this respect—unlike most plants it makes prodigious quantities of jasmonic acid upon wounding). Nevertheless, jasmonic acid does turn up here and there in angiosperms and in some gymnosperms, as well as in some ferns and fern relatives. Unlike OPDA, jasmonic acid and its derivatives have yet to be observed in bryophytes (Stumpe et al., 2010), plants that nonetheless appears to be sensitive to jasmonic acid (Ponce de León et al., 2012). Also, jasmonates are not yet reported in lycophytes. This alone cannot tell us how old jasmonic acid signalling is in plants, but it at least reveals that some early fern relatives such as horsetails (*Equisetum* sp.) can make jasmonate (Dathe et al., 1989). Moreover, common horsetails (*Equisetum arvense*) are known to be highly responsive to jasmonate treatment which, at low micromolar concentrations, inhibits the growth of the tiny gametophyte and can also suppress the formation of the sporophytic shoots that emerge from it (Kuriyama et al., 1993). Furthermore, the concentrations of jasmonate needed to elicit these growth-inhibiting effects in *Equisetum* are in the same range as those that inhibit growth in angiosperms. So together, the reports on jasmonic acid presence and biological activity in *Equisetum* push back jasmonate production and responses, at least for the time being, to this group of early fern relatives. This is long before the origin of dinosaurs.

The horsetail lineage is thought to date back at least 354 million years to the mid-to-late Carboniferous period, long before the radiation of modern ferns (Schneider et al., 2004). What did this early fern-like group have that lycophytes, often dominant components of Carboniferous flora, lacked? One structure is the leaf (as opposed to the microphylls borne by lycophytes). But the early ferns had other features not present in lycophytes, so the appearance of leaves is only one of the advances that would correlate with jasmonate signalling—the same could be said of true roots in ferns and their allies. Moreover, jasmonate appears to affect growth of both the sporophytic and the gametophytic life stages of *Equisetum* and both these life stages are affected by jasmonates in flowering plants. This opens up many possible scenarios for the evolution of jasmonate signalling. If it originated as a means of controlling defence induction it could have arisen to protect the sporophyte (the adult fern) or, equally, it could have played a role in protecting delicate fern gametophytes. In any case, an appearance for jasmonic acid and jasmonic acid signalling in the Carboniferous (or perhaps even earlier) begs the question of what types of plant-eating organisms were abundant in this epoch?

Detritivory versus herbivory

Increasing evidence indicates that through much of the Paleozoic era (that is from roughly 540 to 250 million years ago), detritivory (feeding on dead tissues including those of plants) had a bigger global impact that did herbivory (Labandeira, 1998). As summarized in Figure 5.9, at least four major herbivore expansions over time have been documented (Labandeira, 2006a), the first being indicated by damage to plant lineages that arose long before the Carboniferous period. A second phase of herbivory occurred near the middle of the Carboniferous period, roughly 325 million years ago. Here, many of the plants that were eaten by herbivores were early pteridophytes: ferns and fern relatives; their herbivores included mites and some insect groups (Labandeira, 1998; Labandeira, 2006b). A third phase of herbivore radiation is thought

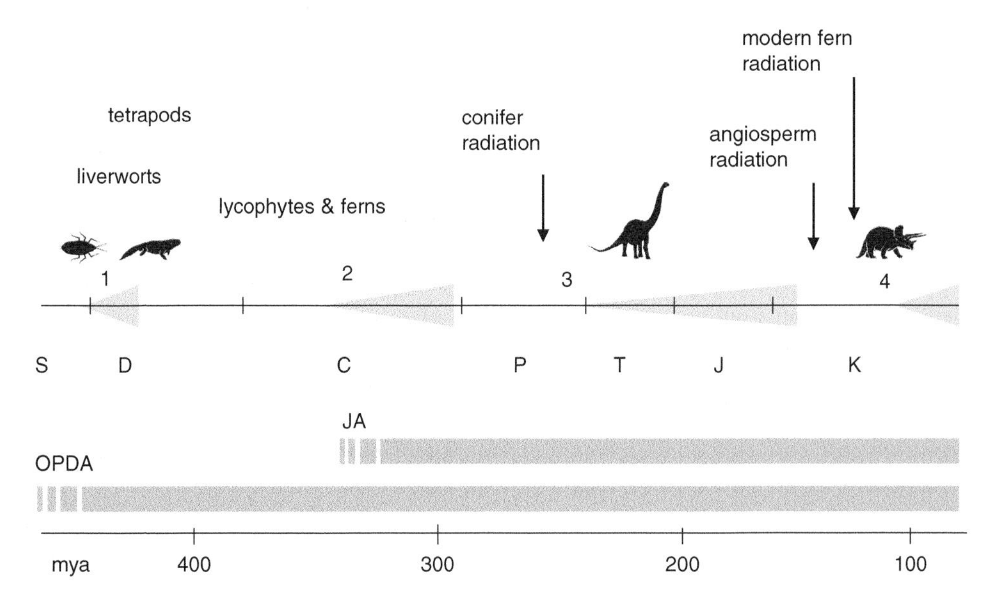

Figure 5.9 Herbivore radiations and the origin of jasmonates. The approximate times of radiations of plant and animal groups are shown in relation to the putative origin of jasmonic acid (JA) and of its precursor 12-oxo-phytodienoic acid (OPDA) in plants. Four major herbivore radiations (Labandeira, 2006a) are indicated as numbers 1–4. Geological periods indicated are: S, Silurian; D, Devonian; C, Carboniferous; P, Permian; T, Triassic; J, Jurassic; K, Cretaceous. The figure presents the hypothesis that JA originated in or before the Carboniferous period. The most ancient plants known both to make and to perceive jasmonates are fern relatives, the horsetails (*Equisetum* spp.).

to have taken place in the Triassic period where beetles started to make their mark as herbivores. Finally, a fourth and most recent phase of herbivore expansion occurred in the Cretaceous period and is where several modern herbivore-rich insect groups radiated—insects such as butterflies, grasshoppers, and hemipterans. In this scheme and from our current knowledge, the emergence of the ability to make and perceive jasmonic acid derivatives correlates roughly with the second herbivore radiation depicted in Figure 5.9. This occurred at a time in the Carboniferous period when herbivore pressure on plants appears to have increased markedly and, it seems, when the proportion of herbivores to detritivores also increased (Labandeira, 2006a, 2006b; DiMichele and Gastaldo, 2008).

At this point, and knowing how strongly the gene products that are under the control of the jasmonate pathway can restrict the growth of herbivores, one can ask whether the reach of this pathway may also extend to affecting the feeding behaviour of detritivores. For this type of experiment there are many available jasmonate mutants in a wide range of plants. The problem is more in choosing a detritivore. Many crustaceans are would-be folivores, some hovering near the limit of being able to consume living leaves. This behaviour can be observed readily on forest floors in the tropics where land crabs can sometimes be seen collecting freshly fallen leaf litter. Moreover, some crabs even clip healthy leaves off plants to bring them down into their burrows (Hicks et al., 1990), and other crabs climb mangrove trees to scrape off and eat cells from the undersides of leaves (Vannini and Ruwa, 1994). These are rather exotic crustaceans, but woodlice, their isopod relatives, are omnipresent wherever the humidity is sufficiently high. These crustaceans are readily available and easily maintained experimental organisms.

In captivity, the common woodlouse (*Porcellio scaber*) cannot sustain itself on healthy leaf tissue even if it is given no other choice. But if this isopod is first starved for two days, then given only a healthy plant to eat, it will nevertheless do its best to try to turn it into food. Starved woodlice will, for example, often sever petioles or stems and the fallen plant parts, especially leaves, are occasionally consumed. No matter how hungry they are, the woodlice need their leaves dead. So for the experiments, *Arabidopsis* plants in which the jasmonate signal pathway was compromised by single mutation in a gene for jasmonic acid biosynthesis were then offered to the woodlice. When the hungry woodlice entered the arena and encountered the wild-type plants they ignored them. But when they encountered jasmonate mutants these plants were immediately consumed alive. Therefore, a single genetic lesion in a plant had converted the crustacean from a detritivore into a herbivore (Farmer and Dubugnon, 2009). This showed that jasmonates can control a major biological transition, that from detritivory to herbivory. Remarkably, this transition required only a mutation in the food source rather than in the animal itself.

What was it that allowed isopod detritivores, organisms that normally feed only on dead tissue, to feed on living jasmonate mutant plants? Obvious candidates were glucosinolates. Brassicaceae leaves typically contain these chemicals; *Arabidopsis* has the aliphatic and indole groups. In simple experiments designed to compare feeding behaviour on wild-type plants and glucosinolate mutants we found that reducing the levels of either class of glucosinolate facilitated woodlouse feeding on otherwise wild-type plants. So part of the reason that the wild-type *Arabidopsis* was not eaten alive was because of its content of glucosinolates. This is not unexpected. Well-defended leaves (e.g. pine needles or holly leaves) decompose slowly in large part because they are highly resistant to detritivory (Grime et al., 1996).

In summary, the evolution of defence chemicals almost certainly preceded the evolution of jasmonate signalling, and this is borne out by the fact that early land plants such as bryophytes and lycophytes have their own defence chemistries (e.g. Rempt and Pohnert, 2010; Ma et al., 2007). One scenario is that as leaves and roots started to appear, so did a refined regulatory system that could co-ordinate defence and growth throughout the plant body. This was the jasmonate pathway, a signal transduction cascade that is strongly activated by cell ingress (Farmer, 1997). In terms of their roles, jasmonates act to maintain defence barriers against opportunistic feeders, totally excluding detritivores and greatly reducing the range of herbivores that can feed on a given plant. The pathway, which occupies a pivotal role in terrestrial ecosystems, helps to control energy distribution between the primary and secondary trophic levels. It has been an evolutionary success in that all living ferns and seed plants use it, and this is underscored by the ways in which herbivores try to suppress its activity.

The suppression of jasmonate signalling by herbivores

The development of any powerful weapon inevitably leads to struggles for its control, and in this respect the jasmonate pathway is no exception. At present there is no evidence that herbivores have learned to corrupt its function directly by producing molecules that would bind to and inactivate the COI1 receptor. Almost inevitably, some of the organisms that attack plants have learned to gain control over the jasmonate pathway and they do this indirectly by activating another signal transduction pathway involved in defence: the salicylate pathway. This latter pathway uses the mediator salicylic acid to control the expression of genes that encode antimicrobial defences and this pathway is responsible for many of the defence responses required to fight off many of the organisms to the left of the dotted line in Figure 1.2. Perhaps through the necessity of economizing defence expenditure, the jasmonic acid and salicylic acid pathways can antagonize each other's activity, so that when one is fully active the other is far less so (Farmer et al., 2003; Koornneef and Pieterse, 2008). But it is this cross-regulation that has permitted some invertebrate herbivores to make a breach in the defence system.

Returning to the fact that herbivores commonly try to manipulate the physiology of the leaves they feed on, when silverleaf whiteflies (*Bremisia tabaci*) are allowed to feed on *Arabidopsis*, these insects introduce salicylic acid-inducing molecules into the phloem to strongly dampen jasmonate responses (Zarate et al., 2007). There is also evidence that something similar takes place at a different phase of the life cycle in chewing insects—that of the egg.

Anything the plant can do to prevent herbivores laying eggs on its leaves will contribute to its survival. Egg mimicry through leaf colour patterning (as sometimes occurs in the Passifloraceae), and other types of leaf patterning that mimic leaf damage, can reduce egg laying, but more general mechanisms are needed in order for the leaf to reduce populations of emerging herbivore larvae. One such mechanism that is probably widespread is the elicitation of volatile production from leaves when eggs are deposited in or on them (Hilker et al., 2002; Colazza et al., 2004). Leaf volatiles can serve as cues for egg-hunting parasitoids, and, like so many other plant defences, the process of volatile release is controlled in part though the jasmonate pathway (Bruinsma and Dicke, 2008). It is therefore very much in the interest of the herbivore to somehow inactivate this pathway. If this is not achieved the tiny larvae may well emerge unscathed from their eggs, but they will immediately have to contend with the formidable range of leaf defences regulated by jasmonate signalling. Together, these defences can slow even the largest insect larva, so they present a considerable barrier.

As a counter strategy, newly deposited large cabbage butterfly (*Pieris brassicae*) eggs and also eggs laid by the generalist Egyptian cotton leafworm (*Spodoptora littoralis*) suppress the activity of the jasmonate pathway by releasing elicitors to stimulate leaves to produce high levels of salicylic acid. (These elicitors presumably mimic endogenous regulators of salicylate production). Invisible to the human eye, this process begins within two days of the eggs being laid and this suppresses jasmonate responses ahead of larval emergence (Little et al., 2007; Bruessow et al., 2010). Then, as they first emerge, the larvae first consume their egg shells before switching to their first meal of leaf tissue in the close vicinity of where they hatched. Here, in the region of the egg, is where the activity of the jasmonate pathway is suppressed most powerfully. Nevertheless, within days, the rapidly growing larvae have to adapt to living on an increasingly hostile leaf in which the residual effects of the eggs on suppressing the jasmonate pathway will soon start to disappear and on which their own activities in damaging the leaf will inevitably stimulate jasmonate signalling. Such invisible battles between eggs and leaves are the norm, but occasionally a plant's reaction can be more violent. This occurs when the leaf recognizes the egg as it would an avirulent pathogen. In this case the tissue immediately below the egg dries, dies, and forces the egg to drop off the leaf (Balbyshev and Lorenzen, 1997). In other cases insect eggs can provoke strong cell division beneath them and this lifts the egg off the surface of the plant to expose it to predators and to desiccation (Doss et al., 2000).

6

Top-down pressures and indirect defences

Direct defences—defence chemicals in resin ducts or laticifers, or spines and stings, etc.—typically function to slow feeding or impede digestion, forcing the herbivore to expend more energy in food procurement. Specialist herbivores, however, have adapted to overcome these defences—although this can still be slow work, and the time taken by a specialist beetle that carefully disarms the pressurized contents of laticifers prior to feeding on a milkweed exposes it to danger from the third trophic level, the herbivore's predators. Such top-down pressures from carnivores are a complement to the plant's own direct defence capacity that puts bottom-up pressure on the herbivore. The direct defences of leaves, the subject of most of this book, are therefore not the complete story. An understanding of plant defence is incomplete without considering the impact of carnivorous animals (Price et al., 1980; Kessler and Heil, 2011).

Plant population remodelling after carnivore removal

Carnivores can manipulate herbivore population structures in complex and often unpredictable ways. An illustration of this has come from a combination of field work and computer modelling of the cyclic population dynamics of collared lemmings (*Dicrostonyx groenlandicus*) in Greenland. The lemming was the sole vertebrate herbivore in the environment but it was prey for four different carnivores: a stoat, the arctic fox, and two bird species, an owl and a skua. Careful study of predation revealed that under most conditions it was the two birds that consumed most lemmings, but removal of any one of the four predators affected the 4–5-year lemming cycles. Whereas removal of the foxes had relatively little effect, removal of the stoat caused lemming population dynamics to be non-cyclical. Finally, the removal of either bird species allowed the lemming populations to escape predator control to the extent that factors other than predation became population-limiting (Gilg et al., 2003). Things change more radically when all major predators are removed. In the total absence of carnivores, herbivores at first flourish but this can lead to massive ecosystem change—above all to plant communities. An elegant demonstration of this came from work in Venezuela where the creation of artificial islands due to a damming project led to such strong habitat restriction that the predators of vertebrate herbivores could not maintain their own populations. The islands, which were created by flooding, simply became too small for animals such as cats and mustelids. In response, populations of vertebrate herbivores prospered (e.g. howler monkeys, as well as some invertebrate herbivores, e.g. leaf-cutter ants), their numbers increasing from 10- to 100-fold depending on the island (Terborgh et al., 2001). At the same time, the density of canopy tree saplings on the small islands dropped to only 20% of that on the large islands. 'Ecological meltdown' had occurred. Rapidly, the plant species composition of ecosystems change as the excess of herbivores selects for strongly herbivore-resistant species. Only those plants with the most performant direct defences will reproduce efficiently and the end result can be seen in many parts of the world where human activity has been too intense. Bottom-up and top-down defences are linked at ecological scales.

In nature, everything seems to be stacked against herbivores. Through each stage of their lives from eggs to adults, insects have to run the gauntlet of attack, and the story is similar for vertebrate herbivores in places where large predators of these animals still exist. Based on this, one could imagine that if the plant could somehow control a predator's behaviour so as to attract it to its prey, it could extend its defence shield by making life even more difficult for the herbivore. Remarkably, the ability to do this is widespread and, for many plants, this is a crucial survival strategy. The first trophic level (plants) has evolved partnerships with organisms in the third trophic level in the joint war against herbivores—the second trophic level. These are the indirect defences of plants. There are many different mechanisms for doing this, but some are used repeatedly. For example, one of the common ways in which plants promote beneficial tritrophic interactions is by providing food and shelter to predatory mites and insects. A second frequently used strategy is to provide information in the form of volatile signals to predators so that they can locate herbivores. Whatever the mechanisms used, indirect defences need to be co-ordinated with direct defences.

Interactions between bottom-up and top-down defence: chemical polymorphism

Before proceeding to look at some of the widespread types of indirect defences it is necessary to examine just how interrelated direct and indirect defence can be. Evidently, a plant's resource investment has to be partitioned between direct defence (such as chemical production) and the maintenance of indirect defence (such as rewarding protective ants with sugar). This has many consequences and, as will be seen in the following example, this can help to explain why many direct defence features of plants are frequently polymorphic. Here the example is provided by the ragwort (*Senecio jacobaea*; Asteraceae), a common plant found in fields and on sand dunes in northwestern Europe. In summer one often finds the black-and-yellow-striped larvae of the cinnabar moth (*Tyria jacobaeae*) on these plants. *Tyria* is a specialist herbivore, but it is not alone on feeding on the plant. An aphid (*Aphis jacobaeae*) is also commonly found on ragwort and although it fills a different feeding niche it is a potential competitor to the cinnabar moth. Unlike its competitor, the cinnabar moth larvae, the aphid has protection in the form of the black garden ant (*Lasius niger*) which farms the aphids for their honeydew. This *Senecio*–*Aphis*–*Tyria*–*Lasius* association can be seen as a 1:2:1 type of interaction where the numbers mean that one plant occupies the first trophic level, two herbivores occupy the second, and one protective organism, an ant, occupies the equivalent of the third trophic level. The fact that there are two herbivores spells trouble for one of them.

Being good farmers, the ants do not accept that other herbivores compete in any way with their aphid herds on the ragwort, so they will harass the *Tyria* larvae, reducing their feeding activity and even controlling their numbers on these plants. However, to complicate things, the ragwort produces a family of nitrogen-containing defence chemicals known as pyrrolizidine alkaloids (named after a conserved part of their chemical skeleton). Crucially, the two herbivores portrayed here display different levels of tolerance to these alkaloids. *Tyria* is able to live on plants with high alkaloid concentrations and it sequesters these compounds for its own defence. By contrast, the aphids are less tolerant of the alkaloids. So in plants where alkaloid concentrations are elevated, aphid infestations will fall and this means that ants will no longer visit the ragwort plants and a powerful indirect defence against *Tyria* will be lost. Therefore, on plants where ants chaperone aphids and in which *Tyria* is also present, it is not advantageous for the plant to accumulate very high levels of alkaloids. In nature, the levels of pyrrolizidine alkaloids in *S. jacobaea* vary at least 10-fold across populations—a polymorphism explained in part by these tritrophic interactions (Vrieling et al., 1991; Hartmann, 1999; Hartmann and Ober, 2008). In summary, plant chemistry can strongly influence tritrophic systems and vice versa. Developing too powerful a deterrent can backfire if it excludes a useful organism.

Ant–plant interactions, extrafloral nectaries, and food bodies

The above example of *Senecio* involves one of the most widespread of all types of protective organisms associated with plants—ants, many of which are sugar-lovers that can rarely resist a meal rich in sucrose or glucose. If the sugar is not taken indirectly from the plant via the honeydew of an aphid or a scale insect, the reward can be provided directly through extrafloral nectaries that have evolved on plants for this purpose (Koptur, 1992; Rico-Gray and Oliveira, 2007). Such foliar ant-attracting nectaries appear to be ancient, having evolved in ferns as diverse as the *Polypodium plebeium* (Polypodiaceae), a Mexican epiphyte, and in the widespread bracken fern (*Pteridium aquilinum*). Extrafloral nectaries are seen more easily on many flowering plants from families as diverse as the Cactaceae or the Zingiberaceae. In North America they can be observed at the leaf–petiole junction on some catalpa trees (e.g. *Catalpa speciosa* and *C. bignoniodes*, Bignoniaceae) and in Europe they are conspicuous on petioles of the wild cherry (*Prunus avium*, Rosaceae; Figure 6.1).

Plant–ant interactions vary in the extent to which each partner depends on the other. If the plant simply feeds the ant, such plants are termed myrmecophiles. In closer partnerships, the ants actually live in plant-provided accommodation (the plants are then termed myrmecophytes) and the structures that house these or other protective organisms are called domatia. Just as there is a great deal of diversity in the structures of extrafloral nectaries, there is a great deal of variability in forms of domatia. At one end of the scale there is little or no apparent modification of the plant to house the ants. For instance, ants can inhabit the internodes of some rattans (Arecaceae) and, whether or not ants are present, the internodes of these

Figure 6.1 Extrafloral nectaries on petioles in wild cherry (*Prunus avium*). (See Plate 11.)

climbing palms look the same. In other cases the plant form is highly modified to house the ants, as is apparent in the swollen stem bases of many tropical and subtropical ant-ferns.

Yet other types of readily visible plant-borne structures, food bodies, can be important to the associations between plants and the organisms that help protect them. Often named after the people who described them, food bodies are generally more common in warm regions than in temperate zones, and this is exemplified by tens of different species of *Cecropia* (Cecropiaceae) trees in the neotropics. These trees feed their stem-dwelling ants with two main types of food bodies: Müllerian bodies at the bases of the petioles of young leaves, and pearl bodies produced on the lower epidermis and on young yeaves. Similarly, many *Macaranga* trees (Euphorbiaceae) in Southeast Asia and Africa house protective ants in their stems and feed them with stipule-borne Beccarian bodies or with similar bodies on stems. These are a few examples among many. But in terms of feeding and housing arrangements as well as the defence contribution of the ants, few tritrophic interactions are as spectacular as those of the ant vachellias of the northern and central neotropics. Whereas humans can handle many ant plants with impunity, there are some that should be avoided.

Ant vachellias—taking tritrophic interactions to the limit

What makes ant vachellias interesting is the extent to which the plant and the insect have coevolved. Many of these associations, like the one to be described, are obligate: neither ant nor plant can survive in nature without the other. The ants need the plant to provide both accommodation (domatia) and food (both food bodies and extrafloral nectaries) while the plant depends on the ants for defence. An excellent example is *Vachellia collinsii* (formerly *Acacia collinsii*; Fabaceae), a common tree found in scrub forests from southern Mexico southwards through Central America to Columbia. No visit to the dry forests of these regions is complete without seeing these small trees, and they are usually easy to find along tracks and in scrubland. Even in the summer heat, any discomfort is soon forgotten when one starts to look at these plants and their protectors. The ants themselves (*Pseudomyrmex* spp.) are tiny, only about 3 mm long, and they live in the hollow interiors of paired stipular thorns, most of which are in the range of 3–5 cm long. Produced without elicitation by the ants, the thorns are found all over the plant including on the main stem. But it is the ants themselves that make the neatly circular entrance holes (Figure 6.2). To do this, they select a thorn pair that has almost completed expansion but that has not yet hardened. Somehow, the ants choose the correct

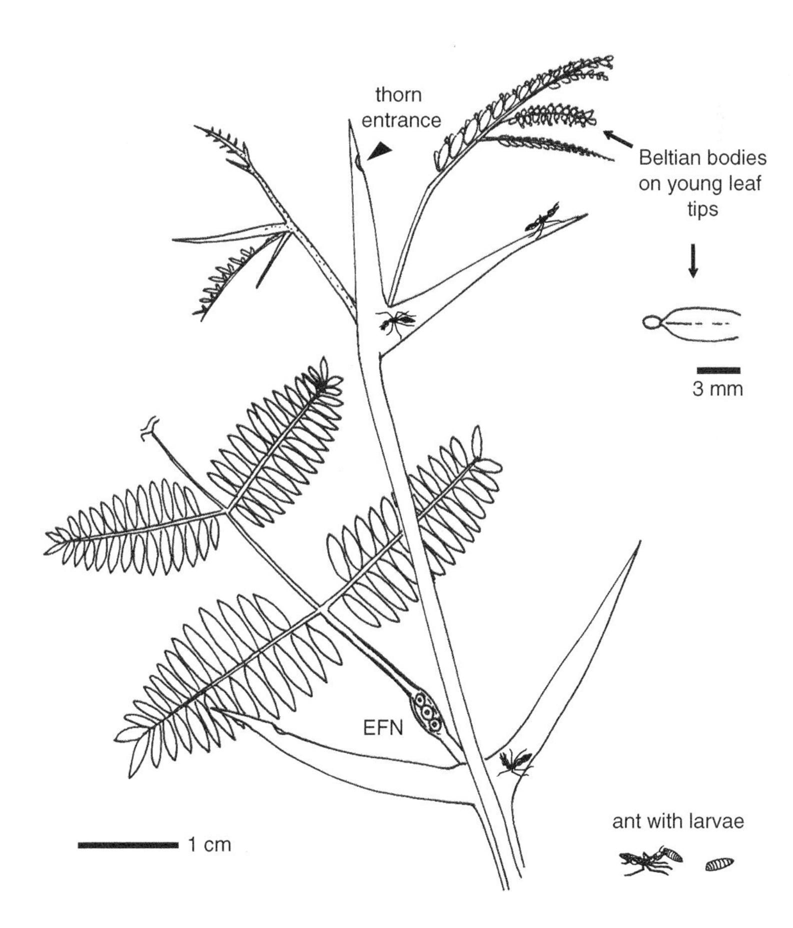

Figure 6.2 The ant plant *Vachellia colinsii* with its *Pseudomyrmex* ants. The image shows the end of a branch. The pinnae that are yet to expand to full size carry Beltian bodies on their tips. Ants clip these off in order to feed larvae that are housed within the hollow stipular thorns. The adults feed themselves from extrafloral nectaries (EFNs). If a caterpillar is placed on the plant, ants pour out of the thorns in order to nip and sting the intruder.

spot in which the tissue will be bitten out and begin to concentrate on cutting into one of the pair of thorns, rarely or never in both. As they do this, other ants go about their business which includes food gathering for their larvae.

A great deal of the day-to-day activity of *Pseudomyrmex* on *V. colinsii* is concentrated at the tips of branches where new growth is occurring. Like any young tissues, these tender red shoots are attractive to a variety of herbivores and they require active protection. Immediately upon expansion of the pinnae, small orange-coloured Beltian bodies are seen on the ends of most of them. These food bodies, however, are transient structures. As the new leaves expand and mature, ants clip the Beltian bodies off and are seen carrying them into the thorn domatia where they will deposit them in small piles to be used to feed the larvae. The adult ants, by contrast, feed on a different, less-protein-rich food—nectar from prominent extrafloral nectaries at the bases of each leaf. As the tree grows, the process of excavating new holes in thorns and bringing food for the young is carried out with typical ant-like expediency, a routine that is only broken when a herbivore intrudes.

Pseudomyrmex is able to detect the smallest intrusion. Within seconds the ants are recruited to whatever touches a leaf, whether it is an insect or a human finger. Despite their small size, they nip and sting fearlessly. If a butterfly larva is placed on the plant it will be attacked swiftly by the ants. Typically, the attacked caterpillar will writhe violently, regurgitate its gut contents, then fall from the plant. Sometimes, however, larvae placed on branches are attacked so violently by large columns of ants that they are immobilized and, through excessive regurgitation, they dehydrate and die. Despite their efficient teamwork and vigilance, the ants and their *Vachellia* do not live in isolation and they do accept certain visitors. Wasps and spiders are often found on the plants and they seem to be tolerated well by the ants since they pose no apparent threat. However, one spider (*Bagheera kiplingi*) has learned to cheat. It is a robber of Beltian bodies, making it the only spider known to display herbivory (Meehan et al., 2009). All this means that only the most highly specialized of herbivores such as *Bagheera* and a ant-resistant few beetle species ever get to taste this plant's leaves. They probably find them to be very palatable. In contrast to many of its relatives in the New World that do not form such close partnerships with ants, the leaves of *V. collinsii* do not have a bitter taste. They are low in the sorts of direct defence compounds found in this branch of the Fabaceae: cyanogenic glucosides.

The *Vachellia*-ant interaction is so intricate that many questions are raised. What, for example, are the plants defended against—vertebrates or invertebrates? Experiments on *V. cornigera* (formerly *Acacia cornigera*) conducted by D.H. Janzen showed that removing colonies of ants from the trees (by spraying with the rapidly-degraded insecticide parathion), usually caused the plant to become badly damaged by herbivorous insects. This classic work demonstrated that defending its host against insect herbivores is clearly a role for *Pseudomyrmex* (Janzen, 1967). But going beyond the invertebrate world, Janzen then extended his observations with *V. cornigera* to mammals. He recorded the activity of a captive female brocket deer (*Mazama americana temama*) feeding on the tree. The deer would feed until stung by the ants, then turn her head away from the branch. This delayed her feeding and she took between three and ten times longer to remove leaves from ant-infested plants than from ant-free plants. Summing up, Janzen stated '... where there are a large number of species to choose from, any factor that causes the deer to turn away from the plant is of significance to the plant, it is unlikely that she would return to the same plant immediately' (Janzen, 1967).

Seeing (and feeling) *Pseudomyrmex* in action raises the question of what sorts of pressures led initially to the evolution of these remarkable ant–plant associations? Of course, the answer to this could be complex since the large thorns on *Vachellia* and many of its relatives presumably first evolved as defences against larger herbivores and were only later co-opted by ants. But having their own armies of ants confers growth advantages since *Pseudomyrmex* sometimes trims unwanted vegetation away from its host (a type of behaviour seen on a number of other ant plants). This not only cuts off easy access routes to many flightless herbivores, but also gives ant vachellias the competitive edge in terms of access to light. My own feeling is that the ant–*Vachellia* associations in the Americas may have evolved under pressure from leaf-cutting ants frequently found in the same habitat. The idea is that ants such as *Pseudomymex* defend their hosts from these other ants. If this were true it would be a case of fighting fire with fire.

The New World ant vachellias provide remarkable examples of indirect leaf defences, but there are also numerous related ant plants in the Old World and notably in tropical Africa. One of the best known of these is the whistling thorn *Vachellia drepanolobium* in eastern Africa (Figure 2.4A). Here the ant domatia are single swollen bladders, otherwise known as pseudogalls, found at the bases of certain thorn pairs. This plant does not produce food bodies nor does it have obvious extrafloral nectaries, so its resident ants therefore either leave the plant to forage, or, as is quite frequently seen, they tend a variety of sucking insects in order to feed indirectly from the host. Depending on the species, the ants on the whistling acacia are sometimes quite laclustre defenders compared with the *Pseudomyrmex* ants in the Americas. But there are ant plants in Africa that could compete with the worst of the New World ant vachellias in terms of their ant's stings. The ants on *Barteria fistulosa* (Passifloraceae) are nearly 1 cm long and have such powerful stings that tying people to the trunks of *Barteria* has been used as a traditional punishment for adultery in Gabon. In summary, few defence-related spectacles in the plant world rival those produced by some of the ant plants of the tropics. When watching ants on these plants attacking caterpillars, one cannot help thinking that similar highly organized but frenetic activity takes place unseen at the molecular level in the herbivore-damaged leaves of all plants. That is, the ants are a visible manifestation of rapid and highly organized direct defence induction in the leaves of other plants.

Mite domatia

Many types of ant domatia can be seen easily from a distance, such as the swollen bases of epiphytic ant ferns in the tropics. But this is seldom the case for the tiny domatia that occur discretely on the undersides of leaves or on petioles of many woody perennials. These are used to house predatory mites that render the leaf the useful services of either attacking herbivores or feeding on potentially pathogenic micro-organisms on the phylloplane. Tree leaves with obvious mite domatia are found relatively infrequently in much of South America, the Horn of Africa, and in drier parts of Australia. By contrast, in temperate regions of North America and Europe, about 40–50% of deciduous trees have mite domatia and the proportion is apparently similarly high for much of Japan and Korea. This does not mean that predatory mites are absent on foliage that lacks clear hiding places for them. The leaves of the pedunculate oak (*Quercus robur*, Fagaceae) in Europe lack domatium-like structures, but by mid-summer their surfaces are well populated, even teeming with mites—predators and herbivores alike. Whereas leaves are conflict zones for organisms such as these, many have a surface topology that has evolved to strongly favour the predators. The two most common types of mite domatia are vein axil trichome tufts, or cavities found on leaves or petioles (Walter 1996; Romero and Benson, 2005). Both of these are usually formed prior to the completion of leaf expansion by the plant alone and without any elicitation or aid by potential occupants.

Spectacular axillary tuft domatia can be seen on the undersides of the leaves of the Macronesian tree *Ocotea foetens* (Lauraceae) in the Canary Islands or on laureltinus (*Viburnum tinum*; Caprifoliaceae), a large shrub found in the wild in northern Mediterranean coastal regions and often grown in gardens (Figure 6.3). The undersides of expanded leaves in this species have prominent axillary domatia within which it is usually easy to observe mites. Laureltinus leaves and their tiny occupants lend themselves well to experimentation focusing on why domatia provide such good homes. When the domatia on these leaves were removed surgically, the number of eggs and nymphs of a mite that feeds by scavenging microbes was reduced on average by ten-fold (Grostal and O'Dowd, 1994). This effect of destroying the mites' accomodation was particularly marked when humidity was

Figure 6.3 Axillary tuft mite domatia on the undersides of leaves in laureltinus (*Viburnum tinum*). (See Plate 12.)

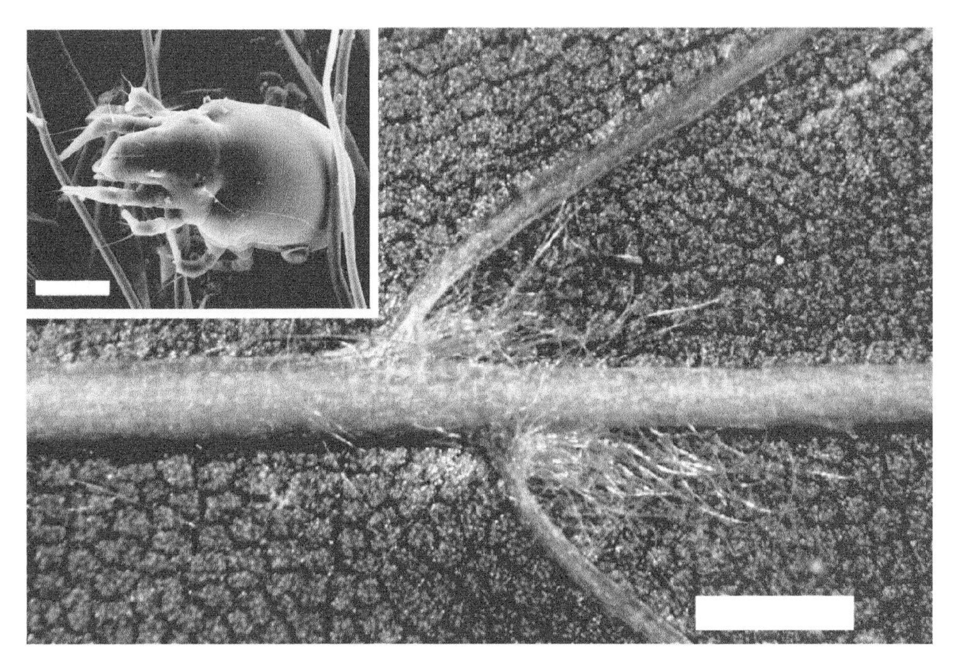

Figure 6.4 Axillary tuft domatia on the undersides of leaves of the common beech (*Fagus sylvatica*). The scale bar for the leaf is 1 mm. The inset shows a mite nymph (probably *Czenspinskia transversostriata*) within a domatium (inset scale bar = 50 µm). Image: A. Muccioli, Lausanne Electron Microscopy Facility.

low, so it seems likely that the microenvironment provided by domatia maintains a high humidity and that this facilitates reproduction. Yet more readily seen examples of trichome tuft domatia are found in the common lime (*Tilia* × *europaea*, Tiliaceae) and some of its relatives and more discrete domatia are seen on the leaves of the cultivated grapevine (*Vitis vinifera*) or the European beech (*Fagus sylvatica*, Fagaceae) (Figure 6.4). By contrast with tuft domatia, cavity domatia are infrequent in temperate regions and are better seen in plants in the subtropics and tropics, e.g. some custard apple relatives (e.g. *Annona muricata*, Annonaceae) and in coffee (*Coffea* spp., Rubiaceae), where some patience is needed to find them.

In summary, both the above examples—domatia and food bodies—are widespread morphological adaptations of plants that facilitate plant–bodyguard partnerships. Both are frequently found in nature and are both amenable to experimental manipulation. However, they are only part of the story. The alternative to such visible manifestations of tritrophic interactions occurs when plants give predators useful information rather than food and lodging. In this case much depends on the predator having a good sense of smell. Not surprisingly—and given the fact that the release of volatiles does not require readily visible structures on leaves—tritrophic systems involving odour-based information transfer were not discovered until relatively recently.

Predator and parasitoid attraction by plant volatiles

Just as a bloodhound follows odours left by the person it is tracking, smaller predators of herbivores can follow scent trails to their prey. The difference is that the dog follows smells produced by its quarry, whereas here the damaged plant has a pivotal role to play—it provides much if not most of the scent. How this works in leaf defence first started to become apparent from studies on mites and the leaves of the Lima bean (*Phaseolus lunatus*, Fabaceae). In this case the herbivore was the two-spotted spider mite (*Tetranychus urticae*), a well-known agricultural pest, and its predator was another mite species (*Phytoseiulus persimilis*) with a known appetite for spider mites. When the herbivorous mite fed on bean leaves it was found to cause the release of volatiles that, even in the absence of the herbivore, faithfully attracted its predator. Therefore, chemicals from damaged leaves were clearly being used to attract a protective organism (Dicke and Sabelis, 1988). At the time, the technical difficulties involved in collecting and analysing small quantities of volatile organic compounds made it difficult to identify compounds released from the plant. Now, the composition of the bouquet is known and it can be chemically reconstructed. We also now know that similar volatile-driven predator behaviour is widespread (Mumm and Dicke, 2010).

Not long after the finding of the tritrophic interaction between the Lima bean and the two mites (one the herbivore, the other the predator), a second type of volatile-based protection system was discovered. This time it involved a grass, maize (*Zea mays*), and herbivores much larger than mites: the larvae of the generalist beet armyworm caterpillar *Spodoptera exigua* (Turlings et al., 1990). When attacked by these caterpillars, the leaves of maize plants (even those that had not been damaged) start to give off a bouquet of volatiles and this develops its full complexity by about 8 h after the initiation of feeding (Turlings et al., 1991, 1998). The volatiles attract predatory and parasitoid insects including the wasp *Cotesia marginiventris*. As adults, *Cotesia* likes to feed on sugar and it readily consumes honeydew from aphids or the nectar produced by flowers. But when they become egg-laydened, the females begin to lose their interest in sugar and their behaviour changes. They fly off to try to locate the distress bouquet given off by an attacked plant since this will help them to locate their caterpillar prey. Once located, the female wasp will inject a single egg into the caterpillar (which is often bigger than she is). Later the egg will hatch and the wasp larva will devour the caterpillar from within.

The key to how volatile signalling works to attract predatory and parasitoid organisms is that leaves of wounded and herbivore-damaged plants typically give off far more volatile organic compounds than do undamaged leaves. Among these are molecules such as 2*E*-hexanal and the compound from which it is derived, 3*Z*-hexenal. These compounds and their relatives including hexenols and hexenyl acetates (Figure 6.5) are central to a group of chemicals known as green leaf volatiles (GLVs)—the chemicals that give freshly cut grass most of its uplifting smell. Simply crushing the leaves of most plants releases at least some of these chemicals, although, depending on the species, the cut-grass smell may be disguised by other dominant leaf chemicals. GLVs emanate to a large extent from chloroplasts and it is likely that the first wafts of GLVs are produced by the breakdown of preoxygenated fatty acids

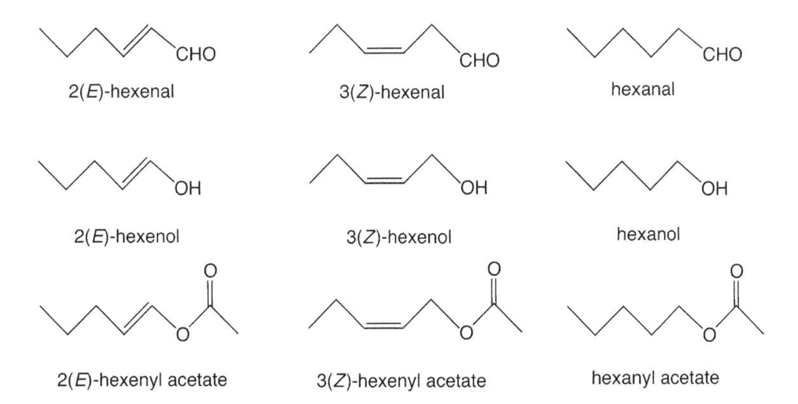

Figure 6.5 Green leaf volatiles (GLVs). These molecules, fragments derived from oxygenated fatty acids, are released when leaves are damaged by feeding herbivores. Most damaged leaves release these compounds as major components of distress bouquets that typically contain other molecules such as mono- and sesquiterpenes and indole. Predatory insects and mites can home in on these bouquets to locate their prey.

on thylakoid lipids. Nevertheless, the enzymatic machinery responsible for GLV synthesis is itself inducible, so more of the compounds are made and released in the hours that follow the initiation of damage. During this time, the composition of the bouquet changes and compounds other than fatty acid derivatives appear. The mixture of volatiles produced by maize is complex and contains about 15 principal compounds that fall into three chemical groups. The major components are GLVs (among which 3*Z*-hexenal dominates), but there are also sesquiterpenes such as E-β-farnesene and the monoterpene alcohol linalool. Additionally, the bouquet contains some phenolic compounds including indole. Ted Turlings the researcher who, in the laboratory of James Tumlinson, discovered much of how this system worked, described this smell to me as 'a mixture of green leaf volatiles and dry hay'. To the entomologist Betty Benrey the bouquet smelt 'like fresh cut grass but stronger'. Could this complex wasp-attracting mixture be reconstructed from individual chemical components as has been achieved for the mite system? The answer is 'not yet'. There appear to be chemically minor components that are yet to be characterized and that play a major role in attracting parasitoids.

How do insect larvae cause such a strong release of volatiles from the plants they feed on? This next step in the research was facilitated by the fact that *S. exigua*, a mid-sized caterpillar, is relatively large compared to a mite. Chemicals originating from the caterpillar could be produced, and the factors that elicited volatile production from maize were tracked down to *Spodoptera*'s oral secretions. Obtaining this fluid (which is often termed 'saliva' or 'regurgitant') is easy. One has only to squeeze the caterpillar gently and it vomits a green liquid. The liquid can then be diluted into water and fed into the transpiration stream of small maize plants that have had their lower stems severed carefully for the purpose. This simple treatment

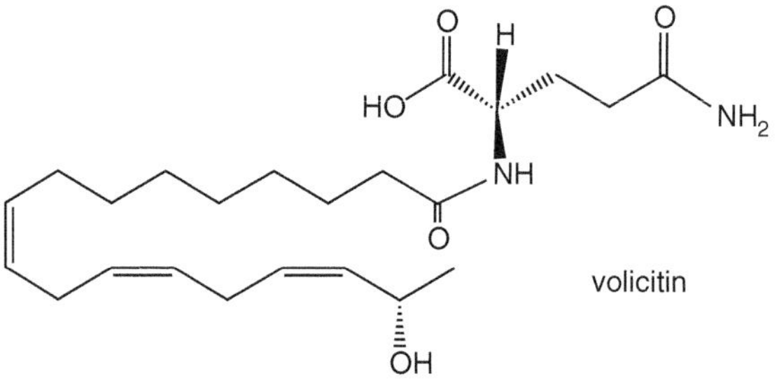

Figure 6.6 Volicitin. Volicitin (17*S*-hydroxylinolenoyl-L-glutamine) is one of several related fatty acid derivatives found in oral secretions from lepidopteran larvae. When applied to plants it strongly enhances the release of the distress bouquet that plants use to attract predators and parasitoids to feeding herbivores.

elicits powerful volatile production that can attract *Cotesia* wasps to the plants, thereby mimicking the feeding of *Spodoptera*. The exciting finalé was that the active component of the regurgitant was eventually traced to a novel compound, 'volicitin', a hydroxy-fatty acid linked to the amino acid L-glutamine (Alborn et al., 1997). Volicitin (Figure 6.6) turned out to be a remarkably active elicitor. Femtomole amounts of this compound triggered volatile release from the maize leaves. Why, then, does the caterpillar give itself away by producing such a powerful elicitor? This is not yet clear and there is still debate about the benefits of producing such a molecule for the caterpillar. But volicitin has surfactant properties and it is possible that it has a useful role during feeding by helping the caterpillar to emulsify lipid components of the plant.

A further question is now raised: is it only voliticin that triggers volatile release or does the damage to plant cells caused by insect feeding have any effect on this process? In other words, what proportion of volatile release is due to crushing of leaf tissue, and how much is due to components such as volicitin released from the herbivore? Mechanical damage alone is reported to be sufficient to induce the release of information-rich volatile bouquets from some plants (Mithöfer et al., 2005). Additionally, plant-derived peptides can powerfully stimulate volatile release (Huffaker et al., 2013). On the other hand, the list of invertebrate-derived substances capable of stimulating volatile release at extremely low levels is growing steadily and volicitin is only one of the insect-produced factors that elicit or modify volatile production in plants (Alborn et al., 2007; Allmann and Baldwin 2010; Hilker and Meiners, 2010). Nevertheless, the fact that the fluid procured from insects is variously termed 'saliva', 'regurgitant', and 'oral secretion' underlines the fact that it may contain components that are not normally present on the mouthparts of the healthy feeding insect: vomit is not saliva. But given the question of where the most powerful elicitors of volatile production come from, the herbivore or the plant, the answer probably lies in the combination of both, as summarized in Figure 6.7. Having first been stimulated by both plant- and herbivore-derived elicitors, leaves then use a relatively conserved chemistry to help to

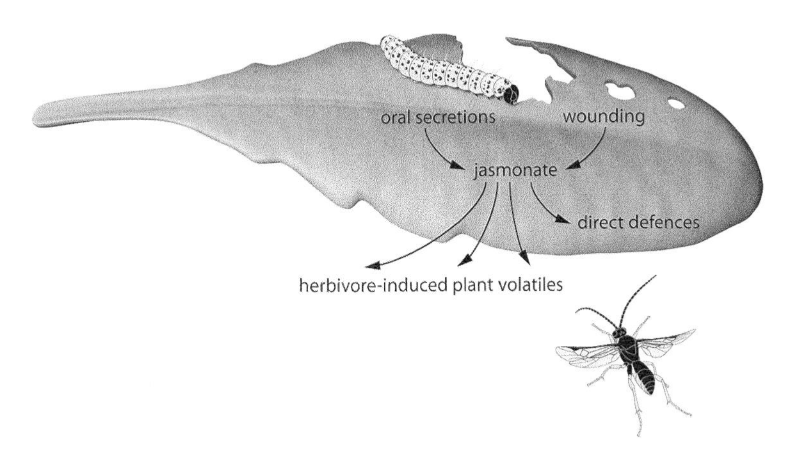

Figure 6.7 Herbivore feeding induces both direct and indirect defences through activation of the jasmonate pathway. A combination of herbivore damage to the leaf and oral secretions triggers the release of a bouquet of volatiles from the plant. These attract parasitoids (e.g. predatory wasps) that will lay an egg in the caterpillar. After Farmer (1997). Image: T. Degen.

attract protective organisms. One can now ask how this process is co-ordinated with the myriad of other active responses of the attacked leaf. Not surprisingly, it appears that the production of the volatile bouquet is controlled in large part through jasmonate signalling: jasmonates couple direct and indirect defences (Boland et al., 1995; Bruinsma and Dicke, 2008; Mumm and Dicke, 2010).

How extensive is the use of plant volatiles to attract predators?

The list of plants known to use volatiles to attract invertebrate bodyguards is growing and includes species as diverse as tobacco and pine (Mumm and Dicke, 2010). Could volatile release from attacked leaves also be used to attract the vertebrate predators of invertebrate herbivores—birds for example? Insectivorous birds are highly visually oriented and, as is easily seen in the forests in Europe, they can home in on leaf damage to locate caterpillars. Blue tits will often fly down to the ground, tilt their heads, and look for holes in leaves made by feeding insects. But work conducted in Northern Finland strongly suggests that volatile release from caterpillar-infested mountain birch (*Betula pubescens*, Betulaceae) attracts these and other insectivorous birds including great tits and willow warblers (Mäntylä et al., 2008). That this might occur makes good sense; after all, many birds have excellent olfaction. On a far more speculative note, perhaps damage-induced volatiles could also attract mammalian insectivores. Often having a highly developed sense of smell, most of these numerous animals are nocturnal and they may not be able to rely solely on sight or hearing to find their prey. Whether this type of plant-guided tritrophic interaction exists would merit investigation.

Kudus and acacias: a cautionary tale

The field of volatile signalling involving plants has more than once been the scene of sensationalism that tells us more about our need to read a good story than about what really occurs in nature. A well-known tale is that, when bitten by kudus, South African acacia trees give off volatile signals that can travel downwind to stimulate defences in undamaged trees. The apparent ability of these trees to release volatile warning signals to induce lethal levels of tannins in their neighbouring congeners was given as a reason for the large-scale mortality in kudus during an extended period of dry weather. Stemming from a report in a park bulletin, this story might have died a natural death had it not been picked up and distributed widely by a larger circulation science publication. With this extra publicity (Hughes, 1990) the story quickly became ingrained in people's minds and remains so even today. I have not been able to trace a peer-reviewed publication detailing these acacia-to-acacia airborne warning signals, but the magazine article mentioned a refereed journal in which the scientist involved later did publish his findings (van Hoven, 1991). The paper does indeed report correlations between tannin levels and the density of kudu populations, but nowhere is there mention of wind-borne signals. As we stand, there is relatively little solid evidence for plant-to-plant defence signalling occurring in the field. Instead, volatile defence signalling within the tissues of plants appears increasingly likely (Farmer, 2001; Kessler and Heil, 2011) and from the point of view of a researcher, this is just as exciting as witnessing the behaviour of parasitoid wasps as they home in on their prey using plant-derived volatiles.

7

Release and escape from herbivory

The direct and indirect defences of plants that are seen today are the result of long evolutionary processes involving the plant and, typically, suites of herbivores. But herbivores sometimes disappear abruptly in the face of human intervention or climate change. This can leave plants with anachronistic defences that are no longer optimal and that will then evolve in new directions to defend the plant against a radically new herbivore (e.g. goats introduced on to a small Pacific island) or the defence will slowly disappear. It should be possible to use anachronistic defences to reconstruct features of extinct herbivores.

Release from herbivory: anachronistic defences

The biological world never quite catches up with the present, and the constant change in the spectrum of herbivores on Earth means that plant defence mechanisms must keep evolving. When a herbivore disappears through extinction, defences that were used against it may remain over generations before they gradually start to decay. A possible example of this is holly (*Ilex aquifolium*; Aquifoliaceae): its lower leaves are prickly whereas higher up the plant the leaves are rounded and lack prickles. Whatever it is or was that attacks or attacked holly, it could not reach up much higher than 3 m off the ground. Today, the prickly leaves of this tree continue to serve it well in the presence of herbivores such as horses. But was this plant once eaten by other macroherbivores that roamed around much of the northern hemisphere until relatively recently? Anachronistic defences are often eroded, but still recognizable, and they both complicate the interpretation of plant populations in terms of defences and also provide windows into the past. However, this phenomenon is not as easily studied in Europe and North America as it is in other parts of the world, and this is due to a combination of human activities and recent glaciations, etc. Instead it is best studied on islands, and in this respect the remaining indigenous and endemic flora of New Zealand still offers a remarkable and very accessible glimpse of the past.

Anachronistic defences in New Zealand

With the exception of an extinct mouse, there is no evidence for indigenous terrestrial mammals on New Zealand's islands. But at least 40% of avian species in New Zealand have disappeared in the last millennium and these included giant ratites called moas (Worthy and

Published 2014 by Oxford University Press.

Holdaway, 2002). These birds put great pressure on the flora: they were the indigenous macroherbivores and some of these birds were very large, weighing up to ~250 kg. Most if not all of the dozen or so species of these birds were exterminated roughly 600 years ago. Nevertheless, pockets of the native flora in New Zealand often contain stunning evidence for the co-existence and co-evolution of moas and plants. Anyone with time and curiosity can find documented and undocumented plant features that relate to past avian herbivory and this is facilitated by good texts on the sometimes unique plant growth forms found on this archipelago (e.g. Dawson, 1988).

In the late 1770s, three centuries after the extermination of the majority if not all of the moas, Europeans arrived and began discovering their large bones, eggs, and even feathers. Although they avidly reconstructed moa skeletons, nobody made the connection to the remarkable defences of the plants that the moas ate. This connection was eventually made by I.A.E. Atkinson and R.M. Greenwood, two researchers who attempted to explain the prevalence of an unusual plant growth form that occurs in more than 50 woody species in the New Zealand flora: divarication. In this growth pattern (Figure 7.1A), most of the sub-branches are at an angle of ≥90° relative to the parent branch. Divaricates are not only found in New Zealand, they are common in parts of Madagascar, and some are found in and near Patagonia, and in Australia, and elsewhere; regions that had or still have giant terrestrial birds (McQueen, 2000; Bond and Silander, 2007). Frequently associated with divarication in New Zealand is heteroblasty—different growth forms and foliage on various parts of the same plant. This growth

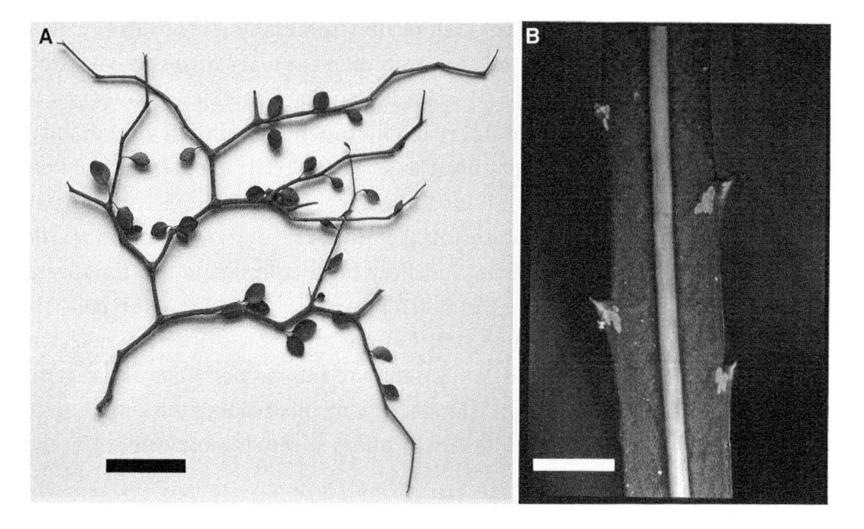

Figure 7.1 Anachronistic defence features in New Zealand *Pseudopanax* (Araliaceae).
(A) Divarication and microphylly in *Pseudopanax anomalus* (herbarium specimen provided by M. Korver, Landcare Research, Lincoln, New Zealand). Scale bar = 2 cm. (B) Pale microleaf patterning on the dark surface of a juvenile leaf of *Pseudopanax crassifolius*. When seen in a dark forest, the bulk of the lamina is almost indistinguishable from the background and the eye is drawn to the leaf decorations. Scale bar = 1 cm.

transition in New Zealand typically occurs between 2 and 3 m above ground. At this level, divarication is suddenly lost, the branches are less acutely angled, and they tend to bear larger leaves. This abrupt change in growth form is remarkable and can be seen at a distance.

But it is the lower branches of the divaricates that are interesting from a defence point of view. Divaricate plants often have small leaves that can be caged within an impressive and impenetrable network of branches. The branches of most of these plants in New Zealand do not look very strong, but trying to break them off can be remarkably difficult, explaining why these plants are sometimes called 'wire plants'. In a classic paper, Greenwood and Atkinson (1977) related divaricate growth to moa browsing. The small-leaved branches low down on heteroblastic divaricate plants would be time-consuming to defoliate and the birds would have had to expend considerable energy in breaking off branches. But above the transition line and out of reach of the birds, the shrubs relax into normal larger-leaved growth forms. It is not conjecture that moas actually ate divaricates: remains of these plants have been found in the preserved guts of these birds (Burrows, 1989). Being divaricate and having small leaves may, then, have been a means of slowing folivory by moas, and this idea was put to the test using the next best thing to live moas: emus and ostriches. Neither bird could swallow the awkward-shaped branches of the plants tested (Madagascan divaricates) and the birds were inefficient at harvesting the small leaves of these plants (Bond et al., 2004; Bond and Silander, 2007). The occurrence of divaricates and other plant defence syndromes in New Zealand is strongly consistent with a major impact of bird-browsing on plants.

Counter-hypotheses to the moa hypothesis have attempted to explain heteroblasty, heterophylly, and divarication as a response to the harsh Pleistocene climate of New Zealand. It has been argued that the concentration of small leaves deep within the branches of some New Zealand divaricates offers protection against the cold (e.g. McGlone and Clarkson, 1993). But why would so many divaricates produce larger leaves at the exposed tops of the plants? In my view the moa defence hypothesis and the climate protection hypothesis are not mutually exclusive, but whereas there are many divaricate plants endemic to New Zealand's two main islands there are few or none that are endemic on the sub-Antarctic islands, on which the climate was as harsh as on the main island but where moas were absent. Finally, divarication, in addition to its association with microphylly, is sometimes combined with other defensive traits. In at least two New Zealand species of divaricating plants, *Discaria toumatou* (Rhamnaceae) and *Melicytus alpinus* (Violaceae), the outermost branches terminate in thorns, making it even more difficult to get at leaves, but these are not the most spectacular examples of physical defences in the New Zealand flora.

Leaf form and patterning in the New Zealand flora

There are many interesting examples of visual markings on the leaves of New Zealand endemics and it is worthwhile considering whether these too could have been selected by moas. One example of this is sometimes seen on the lower leaves of young lancewood plants (*Pseudopanax crassifolius*; Araliaceae). On dark forest floors the pale-coloured midvein and pale green microleaf markings are far more noticeable than the dark leaf blade itself (Figure 7.1B). Some other New Zealand plants play dead, for instance the daisy *Helichrysum depressum* (Asteraceae) and the sprawling shrub *Muehlenbeckia ephedroides* (Polygonaceae). In addition, the juvenile forms of several divaricate shrubs such as *Corokia cotoneaster* (Cornaceae) have dead-looking outer twigs upon which are small leaves. Close inspection of these leaves shows that their petioles (and sometimes the base of the lamina) have the same twig-like colour, making the leaves appear smaller than they really are. It seems, therefore, that a variety of leaf markings on New Zealand plants are there to make leaves appear smaller, that is, to resemble the microphylly found in many of the island's divaricates. If visual trickery such as this exists in the absence of overall changes in leaf form, one may ask if there are real changes in leaf form that might be related to defence against moas. Clues may be found from the way other unrelated birds exert selective pressure on their food.

Birds tend to be highly visually guided when feeding, and insectivorous birds such as jays (*Cyanocitta* spp.) in North America drive the selection of cryptic and polymorphic populations of the moths they eat (Bond and Kamil, 2002). Could extinct (and extant) birds in New Zealand have done similar things with the flora? The answer to this seems to be 'yes', and there are remarkable examples of leaf polymorphism where the moas lived. Examples of this reach their zenith in climbing plants such as the native jasmine (*Parsonsia heterophylla*, Apocynaceae) in which leaves on the same individual can range from elongate to ovate through many bizarrely shaped intermediate forms. But the most convincing of all of New Zealand's relict defences must surely be the thorns on spear grasses as well as the remarkable ways in which their seedlings can mimic the forms of other plants.

Spininess in New Zealand: the spear grasses

If it was only the ongaonga nettle (*Urtica ferox*, Urticaceae; Figure 3.4) on New Zealand that had dangerously powerful relict defences we would still have to ask which herbivores had selected for these stings. But the ongaonga is only one of New Zealand's highly defended plants. A much bigger group are the spear grasses (*Aciphylla* spp., Apiaceae), a group of more than 30 species all but two of which are endemic to New Zealand and its offshore islands, the others being found in southeastern Australia. Modern members of the genus on the New Zealand mainland have a strong tendency to spininess. First, the rosette leaves are often sharp-tipped and are arrayed in a hemispherical pattern with some lying parallel to the ground, protecting the lower stem. Even more impressive are the thorns on the inflorescence stems. On some species (e.g. *A. scott-thompsonii*) these can exceed 40 cm in length and are displayed at head level, so that one has to be extremely wary when close to these plants. These thorns on *Aciphylla* (Figure 7.2) evolved so that moas that tried feeding on flowers and seeds would have risked having their eyes punctured.

The tendency to spininess often correlates with seed size in *Aciphylla*: the larger the seeds the better the defence. Only where these birds existed is *Aciphylla* heavily armoured with inflorescence stem thorns. This holds true for species on New Zealand but breaks down for the Chatham Islands species *A. dieffenbachii*. Moas were absent from these islands. Moreover, in contrast to the heavily defended spear grasses of New Zealand's mainland, the two

Figure 7.2 Spiny New Zealand spear grasses (*Aciphylla* spp., Apiaceae).
(A) *Aciphylla subflabellata*, one of many heavily defended spear grasses in New Zealand, has seedlings that mimic dominant native grasses. Scale bar $\approx$ 20 cm. (B) *Aciphylla scott-thompsonii*, a plant that grows to $\geq$3 m in height and that can be dangerous to humans. These and other spear grasses evolved eye-piercing thorns to defend reproductive organs from moa predation.

Australian *Aciphylla* species, both found at relatively high elevation, lack large thorns. Finally, and in addition to their heavy armoury, the seedlings of *Aciphylla* sometimes resemble the dominant or co-dominant plants with which they grow. *A. subflabellata* seedlings resemble the stiff and narrow leaves of the endemic tussock grass *Festuca novae-zelandiae* and those of *A. aurea* can be hard to find near tussocks of the grass *Chionochloa macra*. The leaves of seedlings of *A. hectorii* have a mottled appearance and resemble those of some *Celmisia* (Asteraceae) species.

In theory, if we had never known of the existence of the moas we could have still learned from the anachronistic defence features of New Zealand's remarkable endemic flora that there were large, visually oriented herbivores that ate leaves and twigs, and could feed up to a height of well over 2 m. With careful work it might even have been possible to describe even more features of these biological ghosts. But what are the limits to this approach? One obvious constraint is time. Over the millennia defence traits would be expected to relax and disappear. How long can visible defence traits endure in the absence of the organisms that selected for them?

How long can anachronistic defences last?

On the other side of the Pacific, far from New Zealand, the Hawaiian archipelago was also devoid of ground-dwelling mammalian folivores until the arrival of humans. Again, like New Zealand, the niches of these vertebrates were filled by birds—flightless geese called moa-nalos. These disappeared around the time of the arrival of humans ~1,600 years ago but the flora of Hawaii, like that of New Zealand, still retains strong indications of the presence of avian herbivores. It has several very prickly plants—most notably in the genus *Cyanea* (Campanulaceae) (Givnish et al., 1994). *Cyanea* on Hawaii appears to have retained its physical defences approximately three times longer than has *Aciphylla* on New Zealand. Going back further in time, we come to the plants of the Americas where many of the large animals that put pressure on the plants have disappeared over the last few thousand years. Nevertheless, there are numerous tree species in the Americas that have large spine-like structures on their trunks. Could these well-defended trunks have prevented now-extinct vertebrates such as giant ground sloths, gomphotheres and glyptodonts from pushing over the trees to feed on their leaves or fruits? Such ideas are considered realistic because many tree species of the Central American lowland forests have large, fleshy fruit containing hefty and well-protected seeds which must have evolved to be dispersed by the large animals that lived in the Pleistocene (Janzen and Martin, 1982). The trees and their fruit remain (today being dispersed by introduced mammals), so perhaps plant defensive traits might have endured over this period too. Indeed, in a colourful review, Janzen used similar arguments in relation to the spininess of part of the Chihuahuan desert flora in Mexico. In that case, the persistent spininess of the vegetation again seems to be related to a Pleistocene mammalian megafauna (Janzen, 1986). This takes us back to the last ice age in the northern hemisphere, in the order of 10,000 years ago. Can we find even older anachronistic defences?

It has been argued convincingly that there are examples of very persistent defence traits that were aimed against animals that disappeared during a relatively short window of human onslaught on the fauna of Australia, starting ~42,000 years ago and lasting ~2,000 years (Johnson, 2006, 2009). For example, and reminiscent of *Aciphylla* in Australia, an acacia called waddywood (*Acacia peuce*, Fabaceae) starts its life as seedlings that resemble grasses. Later, as it develops it shows strong heteroblasty with its juvenile phyllodes on smaller plants being spiny. The adult-phase phyllodes are, by contrast, larger, flatter, and softer. Similarly there are a few Australian shrubs that, like many New Zealand plants, have a divaricating branching structure that has been linked to defence against birds. These plants include bullwaddy (*Macropteranthes keckwickii*, Combretaceae) and the juvenile form of scrub leopardwood (*Flindersia dissosperma*, Rutaceae) (Johnson, 2006). Together, these intriguing examples could mean that anachronistic physical defences can, at least in some cases, persist for at least 40,000 years. But these are, perhaps, near the upper limits so far established for readily visible anachronistic defence persistence. Beyond this, things are far less clear.

The possibility that plants still bear defences against dinosaurs must therefore be treated with caution. Most extant species of plant evolved long after these creatures disappeared. Nevertheless, it is possible that certain architectural traits in plants that were selected very long ago have been retained to the present. Perhaps the ability of many angiosperm trees to regenerate lateral shoots when the tree is severed reflects a distant past adaptation to large herbivores that were capable of pushing the trees over or severely damaging their trunks. This feature is notably rare in gymnosperm trees—has it been lost or did it never evolve? Importantly, anachronistic defences might go beyond obvious phenotypical attributes such as thorns and stings. These 'defences' might also include the ability to tolerate attack and they might even explain certain other physiological adaptations of plants.

To look into this further it is necessary to examine another phenomenon that, like release from herbivory, is not a classical defence. This, instead, is escape from herbivory. A plant that has escaped a major herbivore by evolving into a new niche will perhaps gradually lose a defence trait that it carried when it cohabited with the herbivore.

Escape from vertebrate folivores

An alternative to adapting to herbivores is to escape them in space. If plants that were once under heavy pressure from vertebrate herbivores were able to escape by evolving into niches that were inaccessible to the herbivores, then perhaps such plants still exist, trapped in such niches even if the herbivores were long extinct. One way to compare this release from herbivory is to regard escape as a spatial phenomenon and release as a temporal one. Here, one can bring in a useful framework known as apparency theory.

Apparency theory

Apparency theory states that the more that a species of plant is accessible as a food source in time or space, the more it would need to invest in defence (Feeny, 1976; Rhoades and Cates, 1976). Although this seems obvious it has many ramifications and, while it is old, apparency theory is good. However, most of the emphasis in the literature on apparency was been placed on plants hiding in time. For example, a plant with a low temporal apparency is one that completes its adult phase in a period of low herbivore density. This could be a plant that germinates and starts development in the autumn, overwinters, and completes its life cycle in the spring before herbivore densities increase. A plant with high temporal apparency could be an evergreen tree—a long-lived plant that bears its leaves throughout the year. It would be under most pressure from invertebrates in summer and perhaps more so from warm-blooded vertebrates in winter. But there has been less literature on spatial apparency—the ability of plants, or parts of plants, to escape from herbivores in space—that is to find niches where herbivore density is low. This is, however, equally important. One question arising is where do leaves escape heavy pressure from vertebrates? Since invertebrate herbivores seem to be able to seek out plants wherever they grow, the following will concentrate on vertebrates—animals that lead to the selection of many of the macroscopic defence features found in plants.

Back-boned herbivores can track down plants with remarkable efficiency. The ability of these animals to reach poorly accessible plant populations can be seen first hand on glacier-surrounded nunataks in the Arctic or, more conveniently, further south in the European Alps. One such nunatak in the Alps is the Jardin du Talèfre, a high-altitude site near the centre of the Mont Blanc massive with spectacular views of the Grandes Jorasses. This site can only be reached by crossing the Talèfre glacier but the Jardin is botanically rich and has, at present, 115 species of flowering plants (Jordan, 2010). Given its isolation and small surface area spanning about 400 m by 700 m, a high elevation (2640–3036 m) and a short growing season, one would expect to find few or no vertebrate herbivores in the Jardin. But a visit in 2009 with the express intention of looking for vertebrates revealed otherwise. There were, firstly, several chamois and a snow hare. We found the scat of a fox and we were informed subsequently that the Jardin is inhabited by field voles and snow voles. It is probable that the Jardin is visited by a fifth vertebrate herbivore, the Alpine ibex. This is typical in mountains everywhere. Going beyond the European Alps, Siberian musk deer and the snow sheep forage high in the mountains of eastern Eurasia, and Dall sheep inhabit steep terrain in northern North America. A similar pattern obtains in deserts. Unlike our surprise in finding so many mammalian herbivores in the Jardin, the presence of invertebrates was expected, and, with few exceptions, plants only escape invertebrate herbivores in environments in which they themselves cannot reproduce efficiently. So, despite its relative inaccessibility, elevation, and small surface area, the Jardin is under significant pressure from both invertebrate and vertebrate herbivores. Where, then, are massive populations of leaves freed from vertebrate pressures?

Relief of pressure on the leaves of trees

The answer to where leaves escape a major group of herbivores is, for most of us, close to our doorsteps. In the vast northern hemisphere forests, leaves above ~2 m from the ground are not reached by ground-based vertebrates and there are remarkably few arboreal vertebrates that feed exclusively on leaves. The problems facing such animals in most cool climates of the north would be multiple. Above all, unless they lived in evergreen trees they would have to migrate to have constant food sources, or they might need to hibernate. However, non-deciduous *Eucalyptus* forests suitable for warm-blooded arboreal folivores exist in Australia, notably in the south east of the continent.

At night in these forests, greater gliders emerge from their dens in old trees to feed exclusively on young foliage, and, as these possums become active, koalas have also started to feed on older, more mature *Eucalyptus* leaves (Moore et al., 2010). The *Eucalyptus* genus dominates vast tracts of Australia, providing a potentially huge niche. However, in many other regions the situation is different. Although Africa's central forests have several colobid monkey species that eat large quantities of tree leaves, they do not have the equivalent of a koala, an arboreal vertebrate that eats the leaves of only one genus of trees. Part of the reason for this might relate to forest diversity. Africa, South America, and Asia are generally more diverse in tree genera. The specializations needed to feed on one family or genus are of little use if related tree species are too widely separated among unrelated trees.

Arboreal vertebrate folivores, rare as they are, are dispersed widely over four continents. All live at a biological extreme and they illustrate how hard it is to both live in trees and be a dedicated folivore. If we consider warm-blooded animals that live and reproduce in trees and use leaves throughout the year as their principal food supply, these animals are only found in forests that produce leaves year-round. Arboreal folivores are represented by three-toed sloths in Central and South America; koalas and greater gliders in eastern Australia; indris, ahavis, and sportive lemurs on Madagascar; mantled howler monkeys in the northern neotropics; proboscis monkeys from coastal Borneo; the hoatzin in northern South America; and possibly also by red tree voles from the eastern USA and tree hyraxes from central and southern Africa. It is hard to imagine a more heterogeneous list of animals: many of the major branches of warm-blooded vertebrates are represented, although about half of these animals are primates. The low energy content of leaves, coupled to the need to have a minimum size to house a specialized digestive system, a constant supply of the right types of leaves, and a warm climate are all factors that help to explain why so few birds and mammals have filled the niche of arboreal folivores (Eisenberg, 1978; Cork and Foley, 1991). Remarkably, the sorts of animals that make the best leaf eaters have not been able to move off the ground and this is part of the reason that the world's great forests are so dense. If lagomorphs and ungulates had successfully taken to the trees our world would be different.

Returning to anachronistic defences, and of relevance to the following chapter, the plant that grows into a tall tree will first have to run the gauntlet of herbivory by vertebrates until it outgrows them. So even if it escapes vertebrate predation in its adult phase, having a good defence strategy in the juvenile phase is necessary. Indeed, in many trees this is very obvious: the seedlings of junipers or olives are spiny whereas the adults are, in general, not. These considerations are needed in examining anachronistic defences that could betray the presence of long-extinct herbivores, even if their bones, shells, or eggs have not yet been found. In theory and without prior knowledge, it should be possible to reconstruct the identity of a lost herbivore by looking at anachronistic defence features of endemic floras. These defences could include thorns or stings, or crypsis, etc., and they could extend to other architectural adaptations. A hunt for lost herbivores that was to include the features of trees for clues should take into account the fact that these plants have to pass through the dangerous juvenile phase to grow beyond the reach of most vertebrates.

8

Escape in space: the cliff trees of Socotra

On continental land masses there are (or at least were) vertebrate herbivores of highly diverse types—those that crawl, climb, swim, or even fly. So there are few plant populations that are completely inaccessible to these animals. By contrast, many islands have never supported a broad spectrum of vertebrate herbivores. In many cases indigenous land mammals were absent and, instead, the processes of colonization and adaptation have often led to vertebrate herbivore niches being filled by non-mammals. Critically, if such island herbivores lacked a particular mode of locomotion then some plants on the island might be inaccessible to the herbivore. For instance, if there were islands that lacked herbivores able to climb but that instead had non-climbing herbivores, then, in theory, plants that adapted to live out of reach of such animals might be able to escape predation. Islands provide the opportunity to look for such plants—especially islands rich in inland cliffs. Of interest in this respect is the Indian Ocean island of Socotra.

The cliff trees of Socotra

Socotra

Lying only 250 km east of the Horn of Africa, and situated roughly equidistantly from the east of Somalia and south of Yemen, Socotra is of interest for a variety of reasons. It was the export of frankincense from trees such as *Boswellia socotrana* (Burseraceae), as well as dragon's blood resin from the tree *Dracaena cinnabari* (Dracaenaceae) that helped to carve the name of the island into human history (Cheung and DeVantier, 2006). But these trees are also part of the biological history of Socotra, an island with a rich and well-documented flora. Indeed, there are more than 800 species of vascular plants on the archipelago of which 37% are considered endemic (Miller and Morris, 2004). In terms of animals, the main island is also of interest for its invertebrates, birds, and reptiles, but it has only one indigenous mammal, a shrew that might have been introduced by humans (Cheung and DeVantier, 2006). What is both surprising and interesting (given its proximity to Africa) is that there is no direct evidence that indigenous vertebrate leaf eaters have ever existed on the island or on its neighbours in the archipelago. The comparatively large size of Socotra can be compared with the surface areas of other Indian Ocean islands that have had, or in some cases still have, indigenous vertebrate herbivores.

Rodrigues (109 km^2) had land tortoises and a large flightless bird, the solitaire. Aldabra (156 km^2) still has giant tortoises. Mauritius (2040 km^2) had both the dodo and land tortoises, and Réunion (2512 km^2) had tortoises. Not far from Socotra, near the southern end of the Red Sea, are the Farasan islands (~686 km^2), home to the Farasan gazelle, a subspecies considered to be endemic to the islands. Socotra is larger than any of these islands, having a surface area of ~3,600 km^2 (roughly the size of Majorca, or of Rhode Island), and yet no bones or egg shells from giant birds or from tortoises have been reported. Similarly, no bones of indigenous herbivorous mammals such as ungulates have been found on Socotra. But the island does still have a rich reptile fauna and an even richer invertebrate fauna. Intriguingly, it has scarabid dung beetles that may be endemic. When considering Socotra's flora from the perspective of zoological history one may ask whether there were ever indigenous vertebrate herbivores on the island. Today there are none; the island has landscapes that have been heavily modified by human activity and intense pressure from livestock—goats in particular (Van Damme and Banfield, 2011).

The question of whether there were vertebrate herbivores on Socotra is not new. J. Kingdon (1989), writing of Socotra's endemic plants, stated: 'It is remarkable that many of these endemics have survived as long as they have ... it is clear that few of the endangered plants have evolved any ability to protect themselves from the onslaughts of camels, sheep and goats.' These sentences capture one of the challenges of trying to analyse defences in the Socotran flora. Much of what can be seen today may have been 'scrambled' by the introductions of domesticated animals that have helped to make many Socotran plants rare, favouring instead certain livestock-resistant species. But Kingdon also raises the question of how so many endemic plants have survived to this day under what is extremely high browsing pressure. With the high degree of endemism in the flora, is it possible that some plants might have remnant physical defences that could give clues as to Socotra's faunal history? While anachronistic plant defences have provided a coherent picture of extinct vertebrate herbivores in New Zealand (Greenwood and Atkinson, 1977), and Hawaii (Givnish et al., 1994), Socotra has not yet yielded to such an analysis.

The difficulties

Reaching higher than 1500 m, Socotra's granitic peaks trap cloud moisture, but the dry seasons are harsh near the coast and on the limestone plateaux. Seasonal shortages of water are only part of the story and the island seems to have passed through periodic drying (Fleitmann et al., 2007). Not surprisingly, succulents such as aloes are abundant on Socotra and the island's iconic species; the remarkable and endemic cucumber tree (*Dendrosicyos socotrana*, Cucurbitaceae) can dry out and its leaves shrink down to white, quiescent, paper-like structures. Many other Socotran plants shed their leaves as the island dries and numerous species show microphylly. Smaller leaves are less easily visible than larger leaves so it can be tempting conclude, perhaps for the wrong reasons, that the microphylly found in some of the Socotran flora was an adaptation to make these plants less visible to herbivores. It may equally be an adaptation for drought survival.

Overbrowsing on Socotra

There have always been livestock on the island within human memory; at present they comprise goats, sheep, camels, and some cattle. Have humans and their animals obliterated many traces of indigenous vertebrate herbivores including anachronistic defences in the indigenous flora? If so, for how long has there been livestock on Socotra? Humans reached Australia about 40,000 years ago and rapidly decimated the fauna. It is therefore conceivable, given its proximity to Africa, that we could have reached Socotra very early on and done the same. A striking vegetation is seen on arrival on the north of the island where the shrubs growing on limestone close to the sea (Figure 8.1) can be divided into two types. Most of the plants in the Euphorbiaceae show relatively little or no signs of damage. Examples of this are the woody shrub *Jatropha unicostata* and the spiny *Euphorbia spiralis* which are left largely undamaged by goats. *Croton socotranus*, also in the Euphorbiaceae, is fed on, but it still grows to natural-looking proportions. This is also the case in highly defended plants from other families, plants such as the snake-like vines *Cissus subaphylla* and *C. hamaderohensis* (Vitaceae). These glaucous, trailing vines have fibrous stems that irritate the tongue severely if bitten. The dominance of the Euphorbiaceae on Socotra might be one consequence of the introduction of

Figure 8.1 Coastal limestone habitat on north central Socotra. The pale-barked woody shrub in the foreground is *Jatropha unicostata* and this plant, ~1.5 m high, shares this habitat with caustic latex-producing spurges including *Euphorbia arbuscula* (Figure 4.18) and *E. spiralis* (Figure 8.3B). Also common in this habitat is another Euphorbiaceae, *Croton socotranus*. Just in the picture, lower right, are some dark shoots of a severely browsed *Commiphora* (Burseraceae). The two-stemmed plant at rear left is a desert rose (*Adenium obesum*, Apocynaceae). (See Plate 13.)

livestock. A similar phenomenon involving different plant species is seen in dry areas of eastern Africa that are heavily browsed.

Intermixed with these dominant plants is a second category of plants. In terms of mass these are less abundant than are members of the Euphorbiaceae; few reach the proportions that they would have if they had grown in the absence of domestic animals. All plants in this category are sculpted by livestock herbivory and they grow in variable and often bizarre shapes. On Socotra the plants showing these growth forms include, to name only several, *Lycium sokotranum* (Solanaceae), *Cryptolepis intricata* (Apocynaceae), *Barleria tetracantha* (Acanthaceae), and trees such as *Commiphora parvifolia* (Burseraceae), some individuals of which remain as stunted clump-like plants struggling to reach their potential as attractive trees several metres high. Intricate branching combined with microphylly is frequently found in endemic Socotran plants and these same species often display the topiary syndrome.

Returning to Kingdon's point that it is remarkable that so much of the Socotran flora has survived until today, I suggest that many of the plants present today were likely to have been tolerant to browsing before humans brought their animals. Is it possible that ancient indigenous herbivores, through selecting for certain plant defences, may have helped to save Socotra's flora? If so, what were these herbivores, and what can we learn about this from relict defences in the endemic flora?

The stings on Tragia

The island has an endemic stinging plant, *Tragia balfouriana* (Euphorbiaceae), a climber and sprawler that can reach higher than 3 m. All leaves except the youngest bear active stings (Figure 8.2A). These stings are most concentrated in leaf edge crenations where they mostly face outwards and upwards relative to the plane of the leaf surface. The flower stems of this

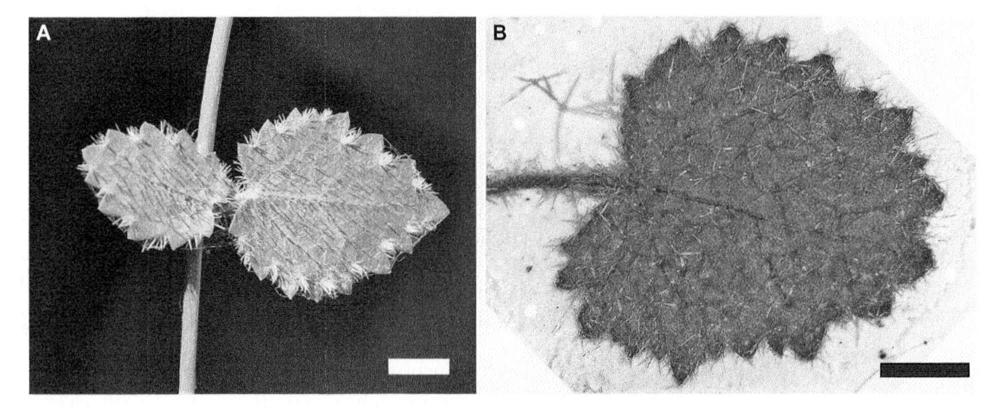

Figure 8.2 Stings and irritant hairs on Socotra. (A) The leaves of *Tragia balfouriana* (Euphorbiaceae) showing axial stinging hairs. Scale bar = 1 cm. (B) A leaf of *Hibiscus noli-tangere* (Malvaceae), a plant that releases highly irritant stellate hairs on contact. Scale bar = 0.2 cm. Note the similarity of leaf forms. (See Plate 14.)

plant also carry dense populations of stings but the vegetative stems are smooth and unarmed. *T. balfouriana*'s stings intensify within the first few minutes of contact, but they are not especially powerful, being of similar intensity to those of the common nettle *Urtica dioica*. Can the sting position give clues as to what *Tragia*'s stings were directed against? The positioning of sting clusters on the edges of leaves does not seem optimal to defend the plant against insects and molluscs that could, in theory, climb the stems and eat leaves beginning from the petiole. But the stings could perhaps have defended the plant against relatively small-mouthed creatures that bit at leaf edges. As a clue to what other plants might be of interest is another feature of *Tragia*: its leaves closely resemble those of several *Hibiscus* species growing in similar habitats on the island.

Irritating Hibiscus

From a defence point of view, *Hibiscus* is clearly one of the more interesting Socotran genera. In fact, the major modern text on plants of the Socotran archipelago warns of the dangers associated with these plants (Miller and Morris, 2004). Socotra has about 16 species of *Hibiscus* (Malvaceae) and it is the leaves of all nine or so endemic species that are notable. An hibiscus with the self-explanatory name of *H. noli-tangere* is typical of these plants. Its leaves (Figure 8.2B) bear star-like hairs with a variable number of arms, most usually three. The largest of these hairs were up to 1.5 mm in width. Smaller stellate hairs of about 1 mm width from same leaf had up to seven arms. To human skin these hairs are as sharp as glass and they are clearly made to detach from the plant when it is disturbed. This is what makes these endemic species unusual. Other *Hibiscus* species on mainland Arabia have almost identical hairs, but they are far less detachable than those of Socotran endemics. Shaking a single leaf—even slightly—loosens a small rain of these structures from the tugged leaf and from leaves above it. The detachability of these hairs in endemic *Hibiscus* may provide clues to the past, and one hypothesis would be that the hairs in these species evolved to be released from the lower sides of leaves to shower down on to a herbivore. This herbivore, if it existed, would probably have had soft body parts in order for these defences to work. Eyes and mouths are the most likely soft parts that would have been targeted.

Endemism and the near absence of large thorns

How many endemic Socotran plants have large thorns (defined here as being thorns of more than 2 cm long) and what could this tell us? *Lycium sokotranum*, a member of the Solanaceae, is an interesting endemic species on Socotra and is, at first sight, potentially useful for analysing the island's faunal past. *Lycium* has thorns of variable length, 1–4 cm long, depending in part on where the plant is growing. Given the variability of thorn length on *Lycium* this plant may be either gaining or losing its thorns with time. However, *L. sokotranum* is the only species in the Solanaceae with its type of morphology on the Socotran archipelago, so it must be used with caution. Highly similar lyciums exist on the Arabian Peninsula and *L. sokotranum* may be a neo-endemic that just happens to be quite goat resistant. Somewhat similar in this respect is *Dichrostachys dehiscens* (Fabaceae), a small tree with thorns up to 2 cm long. This plant is endemic but is again the only representative of its genus and therefore of limited use in the analysis. Are there more informative genera from other families where a range of species has thorns?

Acacias offer another possibility to study thorns. Among the island's acacias it is noteworthy that an endemic species (*Vachellia pennivenia*, Fabaceae) that often grows near *Lycium* typically has only tiny thorns at most about 2 mm long, and if one wishes to find spiny acacias on Socotra then one has to look at non-endemic species. Species such as *Vachellia nilotica* and *Vachellia edgeworthii*, both of which grow on the Horn of Africa, are spiny; endemic Socotran acacias are not. A further indication that large thorns on trees have not been selected for, or maintained on, Socotra is the fact that several other groups of plants can be very spiny on the mainland while their island counterparts are not. This is the case for the myrrh tree *Commiphora myrrh* (Burseraceae) that grows on the Arabian Peninsula and in the Horn of Africa, but not on Socotra. This tree has, by the definition used here, large thorns. Socotra has five *Commiphora* species, four of which are endemic. Three of these endemics are not spiny and one, *C. socotrana*, is only somewhat spinescent. The fifth, non-endemic *Commiphora*, *C. kua*, is far more spiny. This example is representative of a bigger picture: Socotra is not an island on which there is much evidence for the evolution of groups of related endemic plants carrying large thorns. The thorns that are found tend to be less than 2 cm long and they are more common on lower vegetation than on the branches of trees.

The Socotran Acanthaceae and their small thorns

Other plant families on Socotra that are rich in endemics are needed to extend the analysis of spininess on the island. One of these was noticed long ago by the botanist Isaac Bayley Balfour (Balfour, 1888). This is the Acanthaceae, a family that has diversified to the degree that, of the approximately 33 species on the island, 22 are endemic. Several genera of these plants show distinctive defence traits such as idioblasts, reinforced prickly leaf edges, spiny bracts, and great tolerance to herbivory. Some of these plants show intricate branching patterns whereby inner leaves can be protected either by outward-pointing stem thorns (e.g. *Justicia rigida*) or, if they have been browsed, by thick tangles of leafless branches (e.g. *Blepharis spiculifolia*). Thorns and prickles occur in several genera and have roughly conserved dimensions. *B. spiculifolia*, for instance, has prickly leaves of various shapes. Some of these, typically measuring about 1.5 cm from base to tip, end in a single prickle. Others bear two points, and the most common leaf has three prickles arranged as a trident. Another unrelated endemic plant, *Leucas spiculifera* (Lamiaceae), has highly similar leaf shapes and similar prickle dimensions.

Barleria is an interesting genus in the Socotran Acanthaceae. *Barleria aculeata* has hardened leaf edges where the white leaf borders of the small leaves are about 1 mm in width and come to between one and five sharp points each about 2 mm long. *Barleria tetracantha* has, as its name suggests, four-pronged thorns that project outwards from the plant and are borne on stalks up to about 1.2 cm long (Figure 8.3A). One pair of thorns is commonly about 4 mm long, the other closer to 3 mm long. *B. tetracantha* almost certainly did not evolve its

Figure 8.3 Examples of thorns on endemic Socotran plants. (A) *Barleria tetracantha* (Acanthaceae). Scale bar = 0.2 cm. (B) *Euphorbia spiralis* (Euphorbiaceae). (See Plate 15.)

spininess on Socotra since related species with similar types of thorn display are found on mainland Arabia and elsewhere (e.g. *B. trispinosa*). Nevertheless, there are additional physically defended endemic Acanthaceae of Socotra. *Neuracanthus aculeatus* has golden-brown bract thorns up to 1 cm long. Moreover, as mentioned above with the example of *Leucas spiculifera*, it is not only plants in the Acanthaceae that have sharp physical defences that fall into a fairly well-defined size category. One further example of this is the endemic composite *Dicoma cana* that has sharply pointed bracts forming rosettes of up to about 1.4 cm in diameter. These cover the cushion-formed plants, making a very effective barrier.

It is likely that all these plants carried and maintained their thorns as defences against one or more species of indigenous terrestrial vertebrate on Socotra. None of the plants mentioned commonly exceeds 1 m in height (most plants seen in the field are smaller than this). Moreover, with the exception of *Lycium*, few of the thorns or prickly leaves on the plants mentioned here attain ~2 cm in total length. The size of the commoner types of spiny structures on Socotra's endemic Acanthaceae, coupled to their proximity to the ground, would make them candidate defences against non-climbing vertebrates. This would also fit in with observations on *Tragia* and on endemic *Hibiscus* species, and also with *Euphorbia spiralis* (Figure 8.3B), the only spiny succulent on the island.

Heterophylly and leaf polymorphism

Are there other clues as to what sorts of animals could have existed on Socotra? One may be heterophylly and some good examples of this exist on the island. Within the Acanthaceae there is *Anisotes diversifolius* with both small and large leaves. Another possible defence-related trait is polymorphism between individuals in the rare plant *Angkalanthus oligophylla*, also in the Acanthaceae. Some individuals of this plant have narrow lanceolate leaves (reaching more than 10 cm in length and less than 0.6 cm wide) whereas others have ovate leaves 2 or 3 cm in diameter, and yet other individuals have intermediate leaf phenotypes. Other examples exist in ground-dwelling *Boswellia* species (Burseraceae) such as *B. elongata*. Here, juvenile plants can have inconspicuous elongated crenate leaves often well over 15 cm long but less than 0.5 cm wide. As the tree grows, pinnate leaves with narrow lanceolate leaflets are produced. Finally, the adult leaves have much broader pinnae often exceeding 2 cm at the base. *Boswellia elongata* and other variable-leaved plants may conceivably have been using a strategy to evade herbivory, and heterophylly has been associated with protection from tortoises (Eskildsen et al., 2004).

In summary, several defence features in the endemic Socotan flora are consistent with the existence of indigenous vertebrate leaf eaters. They are, first, small thorns and prickles in the order of 0.5 cm long or, when grouped, with overall display dimensions that generally do not span much more than 1.5 cm in diameter. These occur quite frequently on plants that tend to reach only about 1.5 m in height. A second, more restricted, defence feature is the easily dislodged irritant hairs from *Hibiscus* in the Malvaceae. A third shared feature of the endemic flora is heterophylly and leaf polymorphism. Although often hard to separate from responses to environmental pressures such as water stress, some striking examples on Socotra would again be consistent with defence against a vertebrate herbivore. Fourth, there is the ability of many plants to tolerate extreme topiary and to reproduce as stunted forms.

Candidate invertebrate herbivores on Socotra

Among these are numerous terrestrial molluscs. Could gastropods such as *Achatinelloides, Socotora, Riebeckia* or *Balfouria* have driven the evolution of the thorns and stings found on the island's endemic plants? Probably not. The thorns on many of the plants in the Acanthaceae do not seem to be positioned in such a way as to offer good protection to leaves against molluscs (e.g. Figure 8.3A). Even the tough leaves such as those of the Socotran cucumber tree are unlikely to resist snail radulas and, outside Socotra, the rough-surfaced leaves of many plants are often eaten by slugs. Finally, terrestrial gastropods tend to be good climbers. Whether or not any potential herbivore on Socotra could climb is an important issue.

Candidate vertebrate herbivores on Socotra

Mammals

Could leaf-eating mammals have existed on Socotra prior to the arrival of their domestic counterparts? One way to look at this is to abandon defence mechanisms temporarily and to look instead at seed dispersal. Doing this, it is immediately apparent that many good examples of seeds or fruits that have evolved to be dispersed by mammals are found on the island. These include the grass *Setaria verticillata*, the composite *Acanthospermum hispidum*, and the amaranth family member *Pupalia lappacea*. But none of these is endemic or even indigenous and, furthermore, their equivalent is not convincingly found in the endemic flora.

Nevertheless, the presence of a large vertebrate herbivore such as an ungulate cannot simply be dismissed because there are islands smaller than Socotra where apparently indigenous ungulates exist. An example is seen in the lower Red Sea where two of the larger Farasan islands are inhabited by the Farasan gazelle (*Gazella gazella* subsp. *farashani*), a remarkable animal that has survived in a dry, low productivity environment and where it has almost certainly faced centuries of pressure from humans.

The dominant indigenous plants where the Farasan gazelle lives comprise a relatively simple flora characterized by the abundance of spiny plants such as the endemic *Euphorbia collenetteae* (Euphorbiaceae), the indigenous and spiny *Vachellia* (*Acacia*) *ehrenbergiana* and *Maytenus* sp. (Celastraceae), the spiny climber *Capparis sinaica* (Capparaceae), and a small-leaved shruby *Commiphora* (Burseraceae) that, protecting its leaves, has a cage of leafless outer branches that commonly appear to have been browsed. The ground plants include *Blepharis ciliaris* (Acanthaceae), a plant with very spiny bracts. The relative spininess of many of the abundant species in this flora exceeds that of much of Socotra. Given the strong association of spiny acacias with ungulate herbivory and the fact that the endemic Socotran acacias essentially lack thorns, it seems unlikely that ungulates were present on Socotra prior to human arrival. The apparent paucity of endemic Socotran plants with mammal-based seed dispersal mechanisms would be consistent with this.

Birds and reptiles

There is a legend about the 'roc' or 'rukh' in Arabia, perhaps even on Socotra. This was a giant flying bird, and its story (Naumkin, 1993; Cheung and DeVantier, 2006) raises questions about the possible presence of folivorous birds on the archipelago. Flightless and largely herbivorous birds existed on Mauritius (the dodo), on Rodrigues (the solitaire), and on Madagascar (the elephant bird). Indeed, the significant number of microphyllous, intricately branched plants on Socotra is somewhat reminiscent of New Zealand's moa-adapted flora (Greenwood and Atkinson 1977). However, in the endemic Socotran flora there are few plants that resemble the right-angled divaricate branching pattern seen frequently in parts of New Zealand and elsewhere, and no giant thorns such as those found in New Zealand spear grasses (*Aciphilla* spp., Apiaceae). If parallels with New Zealand or Hawaii are considered, there is no evidence that a large folivorous bird existed on Socotra. Although it is difficult to rule out a small herbivorous bird as an indigenous folivore, there seems to be no precedent for this elsewhere in the Indian Ocean. This now leaves reptilian herbivores for consideration.

Lizards that specialize in leaf eating are unknown on Indian Ocean islands and none is known to have placed a strong selection pressure on leaves anywhere near Socotra. Furthermore, most lizards can climb, but the low-to-the-ground defences of plants in the Acanthaceae, and the possibility that the irritant hairs of endemic hibiscus evolved to fall into a ground-based herbivore's eyes, might suggest that Socotra's putative herbivores were ground-dwelling. Tortoises are therefore the primary suspects. This is part of what the anonymous writer of the *Periplus* wrote on the subject nearly 2000 years ago, in the first century AD: 'The island yields tortoise shell, the genuine, the land, the light-coloured, in great quantity and distinguished by their large shields, and also the mountain variety with an extremely thick shell ...' (Casson, 1989). This writing is sometimes interpreted as imaginative (Naumkin, 1993) and it is possible (although unlikely) that the writer of the *Periplus* confused Socotra with another island (Cheung and DeVantier, 2006). However, it fits well with the defence features of the remaining endemic flora. These include thorns and irritant hairs and such defences might work well against tortoises, animals that appear to tolerate certain leaf defence chemicals (Mafli et al., 2012).

Further south in the Indian Ocean, on Aldabra (Seychelles), tortoises are the dominant herbivores. Many herbs growing on coralline limestone on Aldabra can only grow to full size in rock fissures, any exposed plants being browsed down to reduced sizes. These tortoises also feed on trees or shrubs including *Commiphora* (Grubb, 1971). Nowadays, on Socotra, trees such as *Commiphora parviflora* can be found as tight, intricately branched, clumps that struggle to break through to become trees. But these plants might already have been eradicated if the island's flora were not pre-adapted to browsing. Without past pressure from the island's indigenous herbivores, this beautiful *Commiphora* might, along with many other plants, have simply disappeared. Land tortoises of one or more species are the best candidates for the herbivores that left traces in the living vegetation of Socotra. However, if tortoises ever existed on Socotra they may have been exterminated long ago by humans. They may have been sought out for food and, in several places, the *Periplus* mentions with great clarity the hunger faced by people on the island. Empty tortoise shells may even have been used as they still are today in the Horn of Africa, to make 'bells' for goats. If land tortoises did exist on Socotra (until bone of shell is discovered this remains a hypothesis) then their presence might explain other botanical legacies on the island. Above all, this would provide a potential explanation for what is arguably the most interesting phenomenon on Socotra: cliff-dwelling frankincense trees.

Boswellia life-styles

All eight currently described frankincense trees (*Boswellia* spp.) on Socotra are endemic, making them potentially useful as indicators of the Socotra's biological past. Remarkably, these trees have chosen two different life-styles. Three species, *Boswellia elongata, B. socotrana,* and *B. ameero* do what most trees do and live rooted on the ground. Four other species (*B. bullata, B. dioscorides, B. popoviana,* and an as-yet unassigned species) live chiefly on cliffs. A fifth species, *B. nana*, appears to be intermediate. The cliff-rooted boswellias have evolved impressive holdfasts that permit them to cling to vertical surfaces (Figure 8.4). What pressures led to the adoption of cliff dwelling?

There are several possible explanations. One is that formerly more dense forests created intense competition for light and that some boswellias adapted to cliff life to escape being shadowed by other plants. Another original and intriguing possibility, kindly communicated to me by A.G. Miller, is that cliffs trap incoming sea-mist and thus provide a moister habitat than does the ground. A third possibility suggested here is that several boswellias speciated on cliffs under pressure from predation by herbivores that were incapable of climbing. On the

Figure 8.4 A cliff-dwelling Socotran *Boswellia*. The tree is *Boswellia bullata* (Burseraceae), ~5 m high. Note the massive holdfast. Photo: L. Banfield, Royal Botanic Garden Edinburgh. (See Plate 16.)

cliffs the trees could not have escaped from gastropods, or from most birds or from most types of reptile—except non-climbing reptiles such as tortoises. Cliffs can offer refugia from many vertebrates and, today, on the Arabian Peninsula they are home to trees such as *Dracaena ombret*, an increasingly threatened species that is surviving on cliff refugia. But these trees do not have the remarkable adaptations seen in Socotra's cliff boswellias. If cliff-dwelling is an escape strategy in these plants, why has it survived so long?

One possible explanation is that this is because living on cliffs has entailed the evolution of complex holdfasts. Once evolved, such traits would be hard to lose—and they would continue to be of use, today, protecting trees (to some extent) from goats. It is possible that several Socotran *Boswellia* species took to the cliffs in a highly effective escape strategy against herbivores that could not climb. The critical life stages would be seedling establishment followed by the juvenile phase. Once big enough, the trees would be out of danger and would not be expected to be radically different in other respects from their ground-dwelling counterparts. Furthermore, the Socotran boswellias have a close relative in the Horn of Africa. This is *Boswellia frereana* which grows on cliffs and rocky slopes in northern near-coastal Somalia. Like the Socotran boswellias, *B. frereana* has also evolved a holdfast (Thulin and Warfa, 1987). This raises the question of how many more examples of this phenomenon there might be elsewhere in the world. There are indeed other trees that occasionally display holdfast-like structures, and one of these is the yew *Taxus baccata*. But have smaller plants that do not require elaborate holdfasts evolved into cliff niches to avoid predation?

On the other side of the world in the Sonoran region of northern Mexico there are many species of prickly pear (*Opuntia*) cactus, plants that are eaten by reptiles including tortoises (Racine and Downhower, 1974). Almost all of these plants are defended by thorns. Moreover, some opuntias also have glochids that, reminiscent of endemic Socotran *Hibiscus*, can be detachable irritants. But there is at least one exception that is found on cliffs. This is the smooth prickly pear (*Opuntia phaeacantha* var. *laevis*) and it is found on vertical cliffs (Phillips and Comus, 2000). This species is a prickly pear without the prickles and is a further example of a plant that may use cliff refugia to escape herbivory. The cliff-dwelling *Boswellia* species on Socotra may be among the world's more extraordinary examples of this phenomenon.

Synthesis

The implacable force of plant growth depends on the fact that leaves allow plants to build high-complexity structures using the simplest ingredients: soil minerals, water, an inorganic gas, and readily available light energy. Benefiting from enormous power to perform complex chemistry, leaves can also devote substantial energy to protecting themselves. However, this is generally not brute-force defence. Instead, it is often subtle, economical and highly targeted.

Wealth and social context

At the basis of its success is the leaf's modest intrinsic value in terms of energy and resources. The fact that leaves are generally not used as primary storage organs means that many herbivores are drawn to richer tissues such as stems, roots and tubers, and fruits and seeds. To reproduce efficiently it is best to be rich, but to avoid being attacked it is best to be poor. Herbivores seek out the best vegetation on which to feed, and this places different plant species in competition with each other; the more nutritious leaves being potentially more vulnerable than their poorer cousins. So leaves that are intrinsically richer in resources (such as nitrogen) than those of neighbouring plants have to be better defended. Simply by being relatively unattractive from a nutritional point of view, plants that are not heavily armed can reduce predation relative to their neighbours. To a large extent, social context determines investment in defence.

Tolerance and escape

No organ is as tolerant of damage as is the leaf. Even when severely pock-marked by herbivores, leaves can continue to function in carbon capture. The ability of the entire plant body to tolerate herbivory is more variable but tolerance can be a key to survival, as is the case in grasses. Many long-lived trees and bushes also tolerate herbivory well and, when severely damaged, grow as if they had been modelled by topiary. Tolerance is the best complement to defence and can sometimes even replace it but an alternative strategy is to escape certain herbivores by evolving the ability to grow in inaccessible places. Doing so can lead to the evolution of uncommon physiological adaptations.

Published 2014 by Oxford University Press.

Defence expenditure is proportional to attack pressure

Opportunity makes the thief and individual plants that have never endured attack invariably maintain a basal level of defences. The undamaged leaf is extremely responsive to herbivory and quickly reinforces its defences upon insult. If attack is limited, then the defences are activated locally, but when attack is severe, defence is activated widely. This is the main scheme of things, whatever the species. The resources available to an attacker are kept at a minimum before attack and reduced further during attack. No defence has evolved to be more powerful than it needs to be.

The time factor in defence

The use of time and energy factors is a key to robust defence. Rather than killing attackers, leaf defences typically slow food procurement and/or digestion. Leaves use a wide variety of strategies to maximize the time it takes for herbivores to extract food from them. Some plants hide their leaves by making them cryptic, increasing the time it takes a folivore to locate them. Unlike taste, and more so than odour which can disperse in the wind, the colour or hue of leaves can be sensed at a distance except on moonless nights. Leaf markings can confuse would-be egg layers or can visually expose herbivores to predators. Leaf colour patterning can probably be used as a warning too. Spines, trichomes and idioblasts slow leaf eaters. They begin a process of defence first aimed at the eyes, noses, mouths and lips of herbivores. Fibre cells and sclereids make feeding difficult and silica deposits or crystals can wear down mandibles and teeth, both reducing the efficiency of feeding and leading to a more efficient induction of leaf defences. Even specialized herbivores are forced to spend time in countering defence barriers. Once food is taken into the mouth molecular defences come into play and successive layers of these defences will operate during the entire process of digestion some targeting the process of secretion and gut muscle function. By using highly specific toxins to interfere with the nervous systems, leaves can alter the behaviour of folivores and plant-produced regulatory molecules can slow or disrupt the development of small herbivores, again increasing their vulnerability to biotic and abiotic stress. The more time during which a herbivore is preoccupied with food procurement, or the more time it takes to develop, the more easily it can be located by its predators. 'The time-factor rules war' (Liddell Hart, 1944, p. 62).

Energetics: leaves and digestive tracts have co-evolved

Good defences minimize the attacker's energy uptake per unit of time. The leaf is a relatively poor energy source and much of what energy is available is in the form of proteins—the folivore's source of amino acids. Defence strategies that, through their elegance, require relatively little investment by the plant, reduce access to the amino acid content of leaves. To do this, many defence proteins and small chemicals have evolved to target the herbivore's digestive tract. Leaves and stomachs have co-evolved.

In seeking or rejecting food, herbivores use a well-developed repertoire of sensory receptors to detect nocive chemicals in leaves. Ideally for both plant and herbivore, the detection of such molecules leads to immediate food rejection. But if leaves are nevertheless ingested, defence chemicals can have front-end effects on physiologically important receptors such as ion channels. The struggle between the leaf and the folivore is also based to a large extent on how much a chemical costs to make and how much it costs to detoxify. Through their back-end effects on the detoxification system chemicals can place a heavy burden on the animal. Furthermore, increasing the cost-effectiveness of the defence, many chemicals and defence proteins have similar effects on both invertebrates and vertebrates because they often target mechanisms conserved in all animals. Where regulatory mechanisms have diverged in the two animal groups, the leaf has sometimes evolved mechanisms to target each of these mechanisms.

Defence inducibility and the jasmonate pathway

The more it is attacked, the more the leaf makes itself difficult to eat. Upon attack there is rediversion of resources to direct defence, or to rewarding predators from the third trophic level with food, accomodation or information. Alternatively, there is tolerance of attack. Each plant species differs in how it allocates resources to its own protection and this is affected by life-stage and season, but the job of resource allocation in defence is effected by the jasmonate pathway. At the onset of attack, defence chemistry changes quickly, and many metabolites are induced, creating a moving chemical horizon. At the same time, some resources are channelled to the safety of hard-to-attack tissues such as major veins. During attack, signals are also sent to regions of cell proliferation. New growth is modified and leaves that emerge on a previously attacked plant often have characteristics that distinguish them from leaves on unscathed plants. Jasmonate-controlled resource allocation changes can be envisaged as an Egyptian pyramid, a tetrahedron with the four facets corresponding to (i) growth and reproduction, (ii) tolerance, (iii) direct defence or (iv) paid protection. Imagine that 10% of the volume of the pyramid is filled with sand that represents the plant's resources. In ideal conditions the growth face of the pyramid is at the bottom; only a little sand touches the other three facets each of which represents a different defence strategy. But during attack, the pyramid is tilted, growth is reduced, and sand flows either to the facets representing tolerance, defence, or protection. The direction of tilting depends both on the plant and on the strategy used by the attacker. The jasmonate pathway controls the speed and direction of tilting of the defence tetrahedron during herbivory. A highly adapted attacker benefits from allowing or even promoting continued plant growth.

There is no single best defence strategy

A consequence of the fact that plants have such diverse lifestyles is that there is no single optimal defence strategy, the best being one that is highly adapted to each species' own life history. The gymnosperms are good examples of plants in which defences have stood the test of time. Their strategy is to 'grow slowly, build strong defences, and minimize being injured by herbivores'. This is the tactic used by some of the most resilient plants on earth, from cycads to pines. Other plants have life histories that, by being brief, minimize the chances of being eaten. For example, their vegetative growth stages may be squeezed into the space of a few months, the time of emergence of their leaves corresponding to periods of low herbivore pressure. Some achieve this by using storage tissues to fuel rapid growth. At another extreme are the grasses with their grazer-adapted architecture and their successful strategy of 'take the punches and recover to reproduce quickly with fast growth'. Other plants, such as many members of the Solanaceae, also associate themselves with mammals but, using chemical or physical defences, repel attacks by these animals efficiently. Using nutrients excreted by the herbivores these plants grow fast and have the additional nitrogen necessary to invest heavily in producing alkaloids that are then used against the very animals that nourish them. This is only one manifestation of the adage that it can be advantageous to stay close to your enemies.

Leaves co-operate with the enemies of their enemies

The organization of leaf defence is elegant. 'In conditions where two nearly equal forces are pitted against one another, sheer force is of no avail, and only art can bring a decisive result' (Liddell Hart, 1944, p. 297). Bottom-up regulation of herbivore numbers based on chemical and physical defences is coupled to top-down defences that use bodyguards to defend the leaf. In the majority of cases this coupling is loose; other than the provision of habitat in which to hunt herbivores and useful clues from damage-induced volatiles, there is no obvious reward provided to the predator. But many plants go beyond this and have evolved exquisite interactions with mites, ants, and other animals that can attack herbivores. These predators are rewarded with permanent accommodation or with carefully managed information on the whereabouts of intruders and attackers. The coupling of direct defences in the leaf and plant-assisted pressure from carnivores is a question of time and energy—energy rewards to bodyguards, reduction of the attacker's access to energy, and prolongation of feeding times. This is the art of leaf defence.

Relevance of plant defences in agriculture and industry

Since terrestrial agriculture is directly and indirectly leaf-based, it is relevant to investigate how a good knowledge of leaf defences could be best exploited. At present, most agriculture is the maximization of plant growth through the combined process of fertilization and the managed exclusion of pathogens, undesirable insects, and wild vertebrates. However, crop growth and crop protection can be viewed as competing processes. That is, by encouraging rapid growth with generous fertilizer treatments we may in some cases unintentionally reduce the ability of plants to defend themselves. The future challenge will be to manage the two processes optimally while at the same time reducing agricultural inputs (pesticides and fertilizers) and this will increasingly demand a deep understanding of growth and defence control through jasmonate signalling.

Related to this, a continuing drive to reduce the burden of pesticide synthesis and distribution will need to be supported with intelligent strategies to exploit the plant's own defence genes at the appropriate phase in the crop life cycle. What has evolved in nature is often what reduces the use of particular plant tissues in agriculture. That is, defence chemicals, proteins, and physical defences are often central determinants of whether or not we and our livestock can eat raw leaves, seeds, or fruits. To increase food availability for humans and livestock, the defence capacities of certain organs will, through engineering of the jasmonate pathway, be either reduced or augmented. The latter strategy might reduce pesticide treatments, and if a broad range of plant defence genes could be activated it would be difficult for herbivores to evolve resistance to all their gene products.

From a human perspective, two of the leaves' valuable assets are its defence chemicals and its protein contents. However, with only a few exceptions, defence chemicals from plants have played limited roles as crop protectants. One reason for this is that many are intrinsically unstable, this instability often being a key factor in their function. In the leaf, such chemicals are stabilized, sequestered, and repaired in an optimal biological environment, but this form of storage is currently expensive to recapitulate outside the plant. It would be worthwhile to learn more about the mechanisms of biologically maintained stability in order to reduce storage costs for pharmaceuticals, etc. Whereas agriculture selects for edible plants, leaves have evolved so as not to be eaten. This is in part due to the presence of defence proteins that target animal and human digestive tracts. Indeed, one of the reasons that plant foods are cooked is to denature and inactivate these proteins. However, something similar is achieved when the activity of the jasmonate pathway is inhibited because this reduces the levels of these proteins. In addition, the down-regulation of the jasmonate pathway should make tissues less fibrous. The manipulation of the jasmonate pathway either by reducing or increasing its activity in specific organs has potential in both crop protection and in increasing food quality for humans and livestock.

The leaf as a source of amino acids

It is the protein content of leaves that offers future potential as a natural reservoir of amino acids. Half of these molecules are essential to humans and livestock—and, in terms of their energy contents, they are potentially valuable nutrients. Amino acids are also useful molecular building blocks in chemical synthesis. Leaves are perhaps the world's greatest untapped future source of these molecules. If the foliage of more species could be used directly, without the intermediate conversion of leaf contents into animal protein, we would approach the lowest cost and most sustainable source of some of the most valuable molecules, produced alongside nearly every other macro- and micronutrient of importance to humans. However, much of the leaf's natural defence capacity is devoted to reducing access to its proteins and this operates in large part under the control of the jasmonate pathway. In the future we may need to unlock more of the huge potential of leaves as sources of reduced nitrogen, and a starting point in this will be to manipulate jasmonate signalling in sophisticated ways. Making a greater number of leaves available as food will be a question of improving existing crops and domesticating additional plants.

How could we identify new plants to cultivate? Historically, human populations were so attached to their natural environments that we knew what was edible and when to harvest it. Less and less of this knowledge is now passed on. However, this may be achievable a priori, based on a thorough appreciation of the interdependence of plant life histories and defence strategies. Such knowledge should, in theory, help in predicting which plants should be most suitable for domestication. Our knowledge of leaf defences must be used artfully.

Glossary

In many cases consensus definitions for cellular structures and cell constituents related to defence have not emerged or are not always adhered to (e.g. spine, thorn, prickle, gum, etc.). Thorn and spine are herein treated as synonyms with preferential use of thorn for plants and spine for animals. Not all the definitions used below are accepted universally and all refer specifically to the context of this book.

Biforine Raphide crystal-containing idioblast from the aroid family (Araceae) that can expel the crystal when leaf tissue is damaged. Best known in *Dieffenbachia*.

Colleter Glandular trichomes that secrete sticky fluids on to the aerial surfaces of plants.

Crystal idioblasts Idioblast cells containing crystalline deposits.

Crystal sand Populations of sand grain-like crystals, often made of calcium oxalate.

Cystoliths Calcium carbonate crystals found in lithocyst cells.

Defence Physical and chemical features of plants that, through direct and indirect effects, reduce herbivory and thereby maintain reproductive fitness.

Deterrent Compound or structure that reduces the ingestion of plant material after the initiation of feeding.

Detritivore Animal that feeds on dead tissues of plants and/or animals.

Dietary fibre Polymeric chemical components of plants that are not efficiently digested by humans. These are divided into soluble fibres (e.g. pectin and polygalacturonate) and insoluble fibres (e.g. cellulose).

Druses Roughly spherical aggregates of smaller crystals usually composed of calcium oxalate.

Emergence Surface structure upon an organ that is derived from both epidermal and subepidermal cells. Stings, for example.

Epiphyll Organism that grows on a leaf. This is most commonly used for bacteria, algae, fungi, bryophytes, and invertebrates.

Exudivore Animals that feed on plant exudates, usually gums or mucilage.

Fibre See Dietary fibre.

Fibre cell Elongated non-conducting cells with thickened, lignified cell walls.

Folivore Animal specialized to eat living leaves. The term is sometimes restricted to vertebrates, but here it is used for invertebrates as well, including those that feed from within leaves.

Generalist herbivore Animal that feeds on multiple species of plants from different taxa.

Glochid Small, detachable, usually barbed structures formed from trichomes in some cacti (notably many *Opuntia* species).

Granivore Seed-eating animal.

Gum Water-soluble polysaccharides derived from cell wall breakdown and found outside cells in cell walls, extracellular spaces and cavities.

Hair Used only to describe the shape and small dimensions of external, visible structures such as trichomes and glochids. This term conveys no information about the origin of such structures.

Herbivore Feeder on living plant tissue (synonymous with the term 'phytophage').

Heteroblasty Different overall growth architectures (such as branching angles) on the same plant. This is commonly associated with juvenile- and adult-phase transitions but can be subtle. The term is used here where differences in the architecture of parts of plants are unmistakably clear. Many heteroblastic plants are also heterophyllous.

Heterophylly The possession of two or more visibly different leaf or phyllode types on the same plant. This can be found on the same branch of some plants, e.g. in the New Zealand plant *Parsonia heterophylla*. Heterophylly is frequently associated with heteroblasty where each major growth form on an individual plant has its associated leaf morphology (e.g. many New Zealand divaricates).

Hook Loose term used to describe the shape of thorns/spines, emergences, or trichomes that cling to other plants, fur, skin, or clothing. These are chiefly used for climbing or seed dispersal, but many may also have roles in defence.

Hydathode The majority of hydathodes are stomate-like pores used for water release (through guttation). Often closely related to stomata, hydathodes lack regulated guard cells.

Idioblast Single cells that are distinctly different from their neighbours, that have no obvious function in gas, water, and salt exchange, and are primarily defence-related. They commonly store defence compounds (or hydrolases that break down glycosides to release defence chemicals) or crystals.

Latex Milky, light-scattering emulsions based on polyisoprenes that are stored within laticifers. Examples include rubber latex.

Laticifer Tube-like arrangement of storage cells for latex.

Lithocysts Idioblasts containing cystoliths.

Mucilage Water-soluble polysaccharides found within specialized cells, e.g. idioblasts and trichomes.

Oleoresin Resins in which the fluid component is composed of terpenes (usually mono- and sesquiterpenes). Oleoresins based on monoterpenes are common in gymnosperms, especially the Pinaceae. Those based on sesquiterpenes are common in the Dipterocarpaceae (especially *Dipterocarpus* spp.) and the Fabaceae (e.g. *Copaifera* in the New World and *Daniellia* in Africa).

Phytoalexin Low molecular mass antimicrobial compound that is induced in plant tissues upon infection by micro-organisms.

Phytoanticipin Low molecular mass compound that is produced constitutively in plant tissues but which can be synthesized in increased levels in response to attack. The original definition was restricted to antimicrobial compounds but has been broadened herein to include constitutively present antiherbivore compounds.

Phytolith Alternatively known as a silica body. These are regularly-shaped silicon oxide (silica) structures that accumulate in vacuoles or cell walls. Grasses can contain silica cells that are more extensively siliconized cells sometimes containing phytoliths.

Polymorphism Chemical or physical difference between individuals of the same species.

Prickle Used here for a sharp, hardened emergence upon an organ. Prickles contain little or no vasculature and leave a shallow scar when broken off.

Prismatic crystals (prisms) Individual rectangular or rhomboid crystals.

Raphide Elongated needle-shaped crystals often found in easily dissociated bundles.

Raphide idioblast Idioblasts containing raphides.

Repellent Compound, structure, or visual pattern that reduces attempts to feed or to lay eggs on plant tissues (cf. Deterrent).

Resin Lipophilic fluid based on terpenes or phenolics. Resins are stored in specialized extracellular defence-related structures such as resin ducts.

Resin duct Extracellular compartment in which resin is stored. Resin ducts are usually surrounded by epithelial secretory cells.

Resistance Ability to survive attack even if this is coupled to a loss of reproductive fitness.

Sclereid Non-conducting, non-elongated cells with lignified secondary thickening of cell walls.

Sclerenchyma Tissue composed of heavily lignified and usually elongated fibre cells that is used for structural support, storage and defence. Also used for tissues made of populations of sclereids.

Specialist herbivore An animal that feeds on one species of plant or one multiple species from one or a few related taxa. Sometimes extended to animals that feed on unrelated plants that contain closely related chemicals.

Spine Sharp-tipped organ or part of an organ containing vascular tissue. Used here as a synonym for thorn.

Spininess Used here to convey the fact that a certain plant organ bears sharp, hardened structures larger than trichomes.

Sting Surface structure capable of delivering a pressurized irritant solution or sharp crystal into the body of a herbivore. Stings are emergences derived from both epidermal (L1) and subepidermal (L2) tissues.

Styloids Elongated individual crystals.

Thorn Sharp-tipped organ or part of an organ containing vascular tissue. Used here as a synonym for spine.

Tolerance At the whole-organism level the ability to withstand damage without apparent loss of reproductive fitness. Tolerance to severe damage is also a characteristic of most leaves.

Trichome Derived exclusively from epidermal cells and protruding above the epidermal pavement, trichomes can range from simple, unicellular, non-glandular structures to highly complex, multicellular, glandular structures. These can be composed of specialized cell types some of which are devoted to the storage of defence chemicals. Some trichomes may become hardened and detachable.

Trophic island Used here to represent a plant species or variety that differs in its nutritional quality (often through having a unique defence chemistry) with respect to related plants. Without specific adaptations, herbivores cannot overcome trophic barriers.

References

Aardema, M.L., Zhen, Y. and Andolfatto, P. (2012) The evolution of cardenolide-resistant forms of Na^+, K^+-ATPase in Danainae butterflies. *Molecular Ecology* **21**, 340–349.

Adio, A.M., Casteel, C.L., De Vos, M., et al. (2011) Biosynthesis and defensive function of Nδ-acetylornithine, a jasmonate-induced *Arabidopsis* metabolite. *Plant Cell* **23**, 3303–3318.

Agrawal, A.A. (2005) Natural selection on common milkweed (*Asclepias syriaca*) by a community of specialized herbivores. *Evolutionary Ecology Research* **7**, 651–667.

Agrawal, A.A. and Konno, K. (2009) Latex: a model for understanding mechanisms, ecology, and evolution of plant defense against herbivory. *Annual Review of Ecology, Evolution and Systematics* **40**, 311–331.

Ågren, J., Danell, K., Elmqvist, T., et al. (1999) Sexual dimorphism and biotic interactions. In: Geber, M.A., Dawson, T.E. and Delph, L.F. (eds), *Gender and Sexual Dimorphism in Flowering Plants*. Springer-Verlag, Berlin, pp. 217–246.

Aharoni, A., Giri, A.P., Deuerlein, S., et al. (2003) Terpenoid metabolism in wild-type and transgenic *Arabidopsis* plants. *Plant Cell* **15**, 2866–2884.

Aide, T.M. (1993) Patterns of leaf development and herbivory in a tropical understory community. *Ecology* **74**, 455–466.

Alborn, H.T., Hansen, T.V., Jones, T.H., et al. (2007) Disulfooxy fatty acids from the American bird grasshopper *Schistocerca americana,* elicitors of plant volatiles. *Proceedings of the National Academy of Sciences of the USA* **104**, 12976–12981.

Alborn, H.T., Turlings, T.C.J., Jones, T.H., et al. (1997) An elicitor of plant volatiles from beet armyworm oral secretion. *Science* **276**, 945–949.

Allmann, S. and Baldwin, I.T. (2010) Insects betray themselves in nature to predators by rapid isomerization of green leaf volatiles. *Science* **329**, 1075–1078.

Anderson, C.B., Griffith, C.R., Rosemond, A.D., et al. (2006) The effects of invasive North American beavers on riparian plant communities in Cape Horn, Chile—do exotic beavers engineer differently in sub-Antarctic ecosystems? *Biological Conservation* **128**, 467–474.

Antonelli, A., Humphreys, A.M., Lee, W.G. and Linder, H.P. (2011) Absence of mammals and the evolution of New Zealand grasses. *Proceedings of the Royal Society B* **278**, 695–701.

Appel, H.M. (1993) Phenolics in ecological interactions: the importance of oxidation. *Journal of Chemical Ecology* **19**, 1521–1552.

Appendino, G., Minassi, A., Pagani, A. and Ech-Chahad, A. (2008) The role of natural products in the ligand deorphanization of TRP channels. *Current Pharmaceutical Design* **14**, 2–17.

Arnott, H.J. and Pautard, F.G.E. (1970) Calcification in plants. In: Schraer, H. (ed.), *Biological Calcification: Cellular and Molecular Aspects*. North-Holland, Amsterdam, pp. 375–446.

Asano, N., Nash, R.J., Molyneux, R.J. and Fleet, G.W.J. (2000) Sugar-mimic glycosidase inhibitors: natural occurrence, biological activity and prospects for therapeutic application. *Tetrahedron: Asymmetry* **11**, 1645–1680.

Asner, G.P. and Archer, S.R. (2010) Livestock and the global carbon cycle. In: Steinfeld H., Mooney H.A., Schneider F. and Neville L.E. (eds), *Livestock in a Changing Landscape: Drivers, Consequences and Responses*. Island Press, Washington DC, pp. 69–82.

Atkinson, I.A.E. and Greenwood, R.M. (1989) Relationships between moas and plants. *New Zealand Journal of Ecology,* **12** (Supplement), 67–96.

Austin, P.J., Suchar, L.A., Robbins, C.T. and Hagerman, A.E. (1989) Tannin-binding proteins in saliva of deer and their absence in saliva of sheep and cattle. *Journal of Chemical Ecology* **15**, 1335–1347.

Balbyshev, N.F. and Lorenzen, J.H. (1997) Hypersensitivity and egg drop: a novel mechanism of host plant resistance to Colorado potato beetle (Coleoptera: Chrysomelidae). *Journal of Economic Entomology* **90**, 652–657.

Balfour, I.B. (1888) *Botany of Socotra. Transactions of the Royal Society of Edinburgh* **31**, pp. 1–446.

Ballaré, C.L. (2009) Illuminated behaviour: phytochrome as a key regulator of light foraging and plant anti-herbivore defence. *Plant Cell and Environment* **32**, 713–725.

Balouet, J.C. and Olson, S.L. (1989) Fossil birds from late quaternary deposits in New Caledonia. *Smithsonian Contributions to Zoology* **469**, 1–38.

Barlow, B.A. (1981) The loranthaceous mistletoes in Australia. In: Keast, A. (ed.), *Ecological Biogeography of Australia.* W. Junk, The Hague, pp. 577–574.

Barlow, B.A. and Wiens, D. (1977) Host-parasite resemblance in Australian mistletoes: the case for cryptic mimicry. *Evolution* **31**, 69–84.

Becerra, J.X., Noge, K. and Venable, D.L. (2009) Macroevolutionary chemical escalation in an ancient plant-herbivore arms race. *Proceedings of the National Academy of Sciences of the USA* **106**, 18062–18066.

Bee, J.N., Tanentzap, A.J., Lee, W.G., et al. (2011) Influence of foliar traits on forage selection by introduced red deer in New Zealand. *Basic and Applied Ecology* **12**, 56–63.

Beerling, D.J. (2005) Leaf evolution: gases, genes and geochemistry. *Annals of Botany* **96**, 345–352.

Beerling, D.J. and Fleming, A.J. (2007) Zimmermann's telome theory of megaphyll leaf evolution: a molecular and cellular critique. *Current Opinion in Plant Biology* **10**, 4–12.

Bell, A.D. (1991) *Plant Form: An Illustrated Guide to Flowering Plant Morphology.* Oxford University Press, Oxford, p. 76.

Berenbaum M. (1995) Phototoxicity of plant secondary metabolites: insect and mammalian perspectives. *Archives of Insect Biochemistry and Physiology* **29**, 119–134.

Berquist, J. and Örlander, G. (1998) Browsing damage by roe deer on Norway spruce seedlings planted in clearcuts of different ages. 2: Effects of seedling vigor. *Forest Ecology and Management* **105**, 295–302.

Beyhl, D., Hörtensteiner, S., Martinoia, E., et al. (2009) The *fou2* mutation in the major vacuolar cation channel TPC1 confers tolerance to inhibitory luminal calcium. *Plant Journal* **58**, 715–723.

Bidart-Bouzat, M.G. and Kliebenstein, D. (2011) An ecological genomic approach challenging the paradigm of differential plant responses to specialist versus generalist insect herbivores. *Oecologia* **167**, 677–689.

Birgersson, B., Alm, U. and Forkman, B. (2001) Colour vision in fallow deer: a behavioural study. *Animal Behaviour* **61**, 367–371.

Björkman, C., Dalin, P. and Ahrne, K. (2008) Leaf trichome responses to herbivory in willows: induction, relaxation and costs. *New Phytologist* **179**, 176–184.

Bohlmann, J. (2008) Insect-induced terpenoid defenses in spruce. In: Schaller, A. (ed.), *Induced Plant Resistance to Herbivory.* Springer, Stuttgart, pp. 173–187.

Boland, W., Hopke, J., Donath, J., et al. (1995) Jasmonic acid and coronatin induce odor production in plants. *Angewandte Chimie* (international edition in English) **34**, 1600–1602.

Bonaventure, G., Gfeller, A., Proebsting, W.M., et al. (2007a) A gain-of-function allele of TPC1 activates oxylipin biogenesis after leaf wounding in Arabidopsis. *Plant Journal* **49**, 889–898.

Bonaventure, G., Gfeller, A., Rodríguez, V.M., et al. (2007b) The *fou2* gain-of-function allele and the wild-type allele of the Two Pore Channel 1 contribute to different extents or by different mechanisms to defense gene expression in Arabidopsis. *Plant Cell & Physiology* **48**, 1775–1789.

Bonaventure. G., VanDoorn, A. and Baldwin, I.T. (2011) Herbivore-associated elicitors: FAC signaling and metabolism. *Trends in Plant Science* **16**, 294–299.

Bond, A.B. and Kamil, A.C. (2002) Visual predators select for crypticity and polymorphism in virtual prey. *Nature* **415**, 609–613.

Bond, W.J., Lee, W.G. and Craine, J.M. (2004) Plant structural defenses against browsing birds: a legacy of New Zealand's extinct moas. *Oikos* **104**, 500–508.

Bond, W.J. and Silander, J.A. (2007) Springs and wire plants: anachronistic defenses against Madagascar's extinct elephant birds. *Proceedings of the Royal Society B* **274**, 1985–1992.

Borgen, B.H., Thangstad, O.P., Ahuja, I., et al. (2010) Removing the mustard oil bomb from seeds: transgenic ablation of myrosin cells in oilseed rape (*Brassica napus*) produces *MINELESS* seeds. *Journal of Experimental Biology* **61**, 1683–1697.

Bose, J.C. (1926) *The Nervous Mechanism of Plants*. Longmans, Green & Co., London, p. 200.

Brader, G., Tas, E. and Palva, E.T. (2001) Jasmonate-dependent induction of indole glucosinolates in *Arabidopsis* by culture filtrates of the non-specific pathogen *Erwinia carotovora*. *Plant Physiology* **126**, 2556–2576.

Bricchi, I.,Leitner, M., Foti, M., Mithöfer, A., et al. (2010) Robotic mechanical wounding (MecWorm) versus herbivore-induced responses: early signaling and volatile emission in Lima bean (*Phaseolus lunatus* L.). *Planta* **232**, 719–729.

Bruessow, F., Gouhier-Darimont, C., Buchala, A., et al. (2010) Insect eggs suppress plant defence against chewing herbivores. *Plant Journal* **62**, 876–885.

Bruinsma, M. and Dicke, M. (2008) Herbivore-induced indirect defense: from induction mechanisms to community ecology. In: Schaller, A. (ed.), *Induced Plant Resistance to Herbivory*. Springer, Stuttgart, pp. 31–60.

Burrows, C.J. (1989) Moa browsing: evidence from the pyramid valley mire. *New Zealand Journal of Ecology* **12** (Supplement), 51–56.

Cahn, M.G. and Harper, J.L. (1976) The biology of the leaf mark polymorphism in *Trifolium repens* L. 2. Evidence for the selection of leaf marks by rumen fistulated sheep. *Heredity* **37**, 327–333.

Cargill, S.M. and Jefferies, R.L. (1984) The effects of grazing by lesser snow geese on the vegetation of a sub-arctic salt marsh. *Journal of Applied Ecology* **21**, 669–686.

Carmona, D., Lajeunesse, M.J. and Johnson, M.T.J. (2011) Evolutionary ecology of plant defences: plant traits that predict resistance to herbivores. *Functional Ecology* **25**, 358–367.

Casson, L. (ed.) (1989) *The Periplus Maris Erythraei: Text with Introduction, Translation and Commentary*. Princeton University Press, Princeton, p. 69.

Caterina, M.J., Schumacher, M.A., Tominaga, M., et al. (1997) The capsaicin receptor: a heat-activated ion channel in the pain pathway. *Nature* **389**, 816–824.

Chauvin, A., Caldelari, D., Wolfender, J-L and Farmer, E. E. (2013) Four 13-lipoxygenases contribute to rapid jasmonate synthesis in wounded *Arabidopsis thaliana* leaves: a role for lipoxygenase 6 in responses to long-distance wound signals. *New Phytologist* **197**, 566–575.

Chazeau, J. (1993) Research on New Caledonian terrestrial fauna: achievements and prospects. *Biodiversity Letters* **1**, 123–129.

Chen, H., Gonzalez-Vigil, E. and Howe, G.A. (2008) Action of plant defensive enzymes in the insect midgut. In: Schaller, A. (ed.), *Induced Plant Resistance to Herbivory*. Springer, Stuttgart, pp. 271–284.

Chernova, T.E. and Gorshkova, T.A. (2007) Biogenesis of plant fibers. *Russian Journal of Developmental Biology* **38**, 221–232.

Cheung, C. and DeVantier, L. (2006) *Socotra—A Natural History of the Islands and their People*. Odyssey Books and Guides. Airphoto International Limited, Hong Kong.

Chini, A., Fonseca, S., Fernández, G., et al. (2007). The JAZ family of repressors is the missing link in jasmonate signalling. *Nature* **448**, 666–671.

Chow, K-S., Wan, K-L., Isa, M.N.M., et al. (2007) Insights into rubber biosynthesis from transcriptome analysis of *Hevea brasiliensis* latex. *Journal of Experimental Botany* **58**, 2429–2440.

Cipollini, D. (2005) Interactive effects of lateral shading and jasmonic acid on morphology, phenology, seed production, and defense traits in *Arabidopsis thaliana*. *International Journal of Plant Sciences* **166**, 955–959.

Colazza, S., McElfresh, J.S. and Millar, J.G. (2004) Identification of volatile synomones, induced by *Nezara viridula* feeding and oviposition on bean spp., that attract the egg parasitoid *Trissolcus basalis*. *Journal of Chemical Ecology* **30**, 945–964.

Coley, P.D. (1983) Herbivory and defensive characteristics of tree species in a lowland tropical forest. *Ecological Monographs* **53**, 209–233.

Coley, P.D. and Aide, T.M. (1991) Comparison of herbivory and plant defenses in temperate and tropical broad-leaved forests. In: Price P.W., Lewinsohn T.M., Fernandes G.W. and Benson W.W. (eds), *Plant-Animal Interactions: Evolutionary Ecology in Tropical and Temperate Regions*, John Wiley, New York, pp. 25–49.

Coley, P.D. and Barone, J.A. (1996) Herbivory and plant defenses in tropical forests. *Annual Review of Ecology and Systematics* **27**, 305–335.

Coley, P.D., Bryant, J.P. and Chapin, F.S. (1985) Resource availability and plant antiherbivore defense. *Science* **230**, 895–899.

Constabel, C.P. and Barbehenn, R. (2008) Defensive roles of polyphenol oxidase in plants. In: Schaller, A. (Ed.) *Induced Plant Resistance to Herbivory* Springer, Stuttgart, pp. 253–269.

Cooper, R.M., Resende, M.L.V., Flood, J., et al. (1996) Detection and cellular localization of elemental sulphur in disease-resistant genotypes of *Theobroma cacao*. *Nature* **379**, 159–162.

Cork, S.J. and Foley, W.J. (1991) Digestive and metabolic strategies of arboreal folivores in relation to chemical defenses in temperate and tropical forests. In: Palo, R.T. and Robbins, C.T. (eds), *Plant Defenses against Mammalian Herbivory*. CRC Press, Boca Raton, pp. 133–166.

Coté, G.G. (2009) Diversity and distribution of idioblasts producing calcium oxalate crystals in *Dieffenbachia seguine* (Aracea). *American Journal of Botany* **96**, 1245–1254.

Coupe, M.D. and Cahill, J.F. Jr (2003) Effects of insects on primary production in temperate herbaceaous communities: a meta analysis. *Ecological Entomology* **28**, 511–521.

Couplan, F. (2009) *Le régal végétal*: plantes sauvages comestibles. Sang et Terre, Paris.

Crawley, M.J. (1983) *Herbivory: The Dynamics of Animal-Plant Interactions*. Blackwell, Oxford.

Crawley, M.J. (1997) Plant-herbivore dynamics. In: Crawley, M.J. (ed), *Plant Ecology*, 2nd edn. Blackwell, Oxford, pp. 401–474.

Cyr, H. and Pace, M.L. (1993) Magnitude and patterns of herbivory in aquatic and terrestrial ecosystems. *Nature* **361**, 148–150.

Dabrowska, P., Freitak, D., Vogel, H., et al. (2009) The phytohormone precursor OPDA is isomerized in the insect gut by a single, specific glutathione transferase. *Proceedings of the National Academy of Sciences of the USA* **106**, 16304–16309.

Dadd, R.H. (1985) Nutrition: organisms. In: Kerkut G.A. and Gilbert L.I. (eds), *Comprehensive Insect Physiology Biochemistry and Pharmacology*. Pergamon Press, Oxford, vol. 4, pp. 331–380.

Dalin, P., Ågren, J., Björkman, C., et al. (2008) Leaf trichome formation and plant resistance to herbivory. In: Schaller, A. (ed.), *Induced Plant Resistance to Herbivory*. Springer, Stuttgart, pp. 89–105.

Das, I. (1996) Folivory and seasonal changes in diet in *Rana hexadactyla* (Anura: Ranidae). *Journal of Zoology* **238**, 785–794.

Dathe, W., Miersch, O. and Schmidt, J. (1989) Occurrence of jasmonic acid, related compounds and abscisic acid in fertile and sterile fronds of three *Equisteum* species. *Biochemie und Physiologie der Pflanzen* **185**, 83–92.

Davidson, D.W., Cook, S.C., Snelling, R.R. and Chua, T.H. (2003) Explaining the abundance of ants in lowland tropical rainforest canopies. *Science* **300**, 969–972.

Davis, M.A., Pritchard, S.G, Boyd, R.S. and Prior, S.A. (2001) Developmental and induced responses of nickel-based and organic defences of the nickel-hyperaccumulating shrub, *Psychotria douarrei*. *New Phytologist* **150**, 49–58.

Davis, W. (1997) *One River: Science, Adventure and Hallucinogenics in the Amazon Basin*. Simon & Schuster, London, pp. 411–449.

Dawson, J. (1988) *Forest Vines to Snow Tussocks. The Story of New Zealand Plants*, reprinted 1993. Victoria University Press, Wellington.

Dearing, M.D., Foley, W.J. and McLean, S. (2005) The influence of plant secondary metabolites on the nutritional ecology of herbivorous terrestrial vertebrates. *Annual Review of Ecology, Evolution and Systematics* **36**, 169–189.

Dearing, M.D. and Schall, J.J. (1992) Testing models of optimal diet assembly by the generalist herbivorous lizard *Cnemidophorus murinus. Ecology* 73, 845–858.

Dearing, M.D., Mangione, A.M. and Karasov, W.H. (2001) Plant secondary compounds as diuretics: an overlooked consequence. *American Zoologist* **41**, 890–901.

De Vos, M., Van Oosten, V.R., Van Poecke, R.M., et al. (2005) Signal signature and transcriptome changes of *Arabidopsis* during pathogen and insect attack. *Molecular Plant-Microbe Interactions* 18, 923–937.

Dicke, M. and Sabelis, M.W. (1988) How plants obtain predatory mites as bodyguards. *Netherlands Journal of Zoology* **38**, 148–165.

Dillon, P.M., Lowrie, S. and Mckey, D. (1983) Disarming the «evil woman»: petiole constriction by a sphingid larva circumvents mechanical defenses of its host plan, *Cnidoscolus urens* (Euphorbiaceae). *Biotropica* **15**, 112–116.

DiMichele, W.A. and Gastaldo, R.A. (2008) Plant paleoecology in deep time. *Annals of the Missouri Botanical Garden* 95, 144–198 (see especially p. 179).

Dobler, S., Dalla, S., Wagschal, V. and Agrawal, A.A. (2012) Community-wide convergent evolution in insect adaptation to toxic cardenolides by substitutions in the Na, K-ATPase. *Proceedings of the National Academy of Sciences of the USA* **109**, 13040–13045.

Doss, R.P., Oliver, J.E., Proebsting, W.M., et al. (2000) Bruchins: insect-derived plant regulators that stimulate neoplasm formation. *Proceedings of the National Academy of Sciences of the USA* **97**, 6218–6223.

Douglas, A.E. (1998) Nutritional interactions in insect-microbial symbioses: aphids and their symbiotic bacteria Buchnera. *Annual Review of Entomology* **43**, 17–37.

Downum, K.R. (1992) Light-activated plant defence. *New Phytologist* **122**, 401–420.

Dussourd, D.E. and Denno, R.F. (1991) Deactivation of plant defense—correspondence between insect behavior and secretory canal architecture. *Ecology* **72**, 1383–1386.

Dussourd, D.E. and Hoyle, A.M. (2000) Poisoned plusiines: toxicity of milkweed latex and cardenolides to some generalist caterpillars. *Chemoecology* **10**, 11–16.

Ehrlich, P.R. and Raven, P. H. (1964) Butterflies and plants: a study in coevolution. *Evolution* **18**, 586–608.

Eisenberg, J.F. (1978) The evolution of arboreal herbivores in the Class Mammalia. In: Montgomery, G.G. (ed.), *The Ecology of Arboreal Folivores*. Smithsonian Institution Press, Washington, DC, pp. 135–152.

Eisner, T. (2003) *For Love of Insects*. Harvard University Press, Belknap.

Eisner, T., Eisner, M. and Hoebeke, E. (1998) When defense backfires: detrimental effect of a plant's protective trichomes on an insect beneficial to the plant. *Proceedings of the National Academy of Sciences of the USA* **98**, 4410–4414.

Endlweber K., Ruess L. and Scheu S. (2009) Collembola switch diet in presence of plant roots thereby functioning as herbivores. *Soil Biology and Biochemistry* **41**, 1151–1154.

Epstein, E. (1999) Silicon. *Annual Review of Plant Physiology and Plant Molecular Biology* **50**, 641–664.

Erbilgin, N., Krokene, P., Christiansen, E., et al. (2006) Exogenous application of methyl jasmonate elicits defenses in Norway spruce (*Picea abies*) and reduces host colonization by the bark beetle *Ips typographus. Oecologia* **148**, 426–436.

Eschler, B.M., Pass, D.M., Willis, R. and Foley, W.J. (2000) Distribution of foliar formylated phloroglucinol derivatives amongst *Eucalyptus* species. *Biochemical Sytematics and Ecology* **28**, 813–824.

Eskildsen, L.I., Olesen, J.M. and Jones, C.G. (2004) Feeding response of the Aldabra giant tortoise (*Geochelone gigantea*) to island plants showing heterophylly. *Journal of Biogeography* **31**, 1785–1790.

Espírito-Santo, M.M and Fernandes, G. W. (2007) How many species of gall-inducing insects are there on earth and where are they? *Annals of the Entomological Society of America* **100**, 95–99.

Estes, R.D. (1992) *The Behavior Guide to African Mammals: Including Hoofed Mammals, Carnivores, Primates*. University of California Press, Berkeley.

Evans, I.A. (1968) The radiomimetic nature of bracken toxin. *Cancer Research* **28**, 2252–2261.

Evans, W.C. (1976) Bracken thiaminase-mediated neurotoxic syndromes. *Botanical Journal of the Linnean Society* **73**, 113–131.

Evert, R.F. (2006) *Esau's Plant Anatomy: Meristems, Cells, and Tissues of the Plant Body—Their Structure, Function, and Development*, 3rd edn. John Wiley, Hoboken, NJ, pp. 191–209.

Farmer, E.E. (1997) New fatty acid-based signals: a lesson from the plant world. *Science* **276**, 912–913.

Farmer, E.E. (2001) Surface-to-air signals. *Nature* **411**, 854–856.

Farmer, E.E., Alméras, E. and Krishnamurthy, V. (2003) Jasmonates and related oxylipins in plant responses to pathogenesis and herbivory. *Current Opinion in Plant Biology* **6**, 372–378.

Farmer, E.E. and Davoine, C. (2007) Reactive electrophile species. *Current Opinion in Plant Biology* **10**, 380–386.

Farmer, E.E. and Dubugnon, L. (2009) Detritivorous crustaceans become herbivores on jasmonate-deficient plants. *Proceedings of the National Academy of Sciences of the USA* **106**, 935–940.

Farmer, E.E. and Ryan C.A. (1990) Interplant communication: airborne methyl jasmonate induces the synthesis of proteinase inhibitors in plant leaves. *Proceedings of the National Academy of Sciences of the USA* **87**, 7713–7716.

Farmer, E.E. and Ryan, C.A. (1992) Octadecanoid precursors of jasmonic acid activate the synthesis of wound-inducible proteinase inhibitors. *Plant Cell* **4**, 129–134.

Federle, W., Maschwitz, U., Fiala, B., et al. (1997) Slippery ant-plants and skilful climbers: selection and protection of specific ant partners by epicuticular wax blooms in *Macaranga* (Euphorbiaceae). *Oecologia* **112**, 217–224.

Feeny, P. (1970) Seasonal changes in oak leaf tannins and nutrients as a cause of spring feeding by winter moth caterpillars. *Ecology* **51**, 565–581.

Feeny, P. (1975) Biochemical coevolution between plants and their insect herbivores. In: Gilbert L.E. and Raven, P.H. (eds), *Coevolution in Plants and Animals*. University of Texas Press, Austin, pp. 3–19.

Feeny, P.P. (1976) Plant apparency and chemical defense. In: Wallace, J.W. and Mansell, R.L. (eds), *Recent Advances in Phytochemistry*. Plenum Press, New York, vol. **10**, 1–40.

Fernandes, G.W. (1990) Hypersensitivity: a neglected plant resistance mechanism against insect herbivores. *Environmental Entomology* **19**, 1173–1182.

Fine, P.V., Mesones, I. and Coley, P.D. (2004) Herbivores promote habitat specialization by trees in the Amazonian forests. *Science* **305**, 663–665.

Fleitmann, D., Burns, S.J.Mangini, A., et al. (2007) Holocene ITCZ and Indian monsoon dynamics recorded in stalagmites from Oman and Yemen (Socotra). *Quaternary Science Reviews* **26**, 170–188.

Foley, W.J. and Moore, B.D. (2005) Plant secondary metabolites and vertebrate herbivores—from physiological regulation to ecosystem function. *Current Opinion in Plant Biology* **8**, 430–435.

Fraenkel, G.S. (1959) The raison d'être of secondary plant substances. *Science* **129**, 1466–1470.

Franceschi, V.R. and Horner, H.T. (1980) Calcium oxalate crystals in plants. *Botanical Review* **46**, 361–427.

Franco, O.L., Rigden, D.J., Melo, F.R. and Grossi-De-Sá, M.F. (2002) Plant alpha-amylase inhibitors and their interaction with insect alpha-amylases. *European Journal of Biochemistry* **269**, 397–412.

Fromm, J. and Lautner, S. (2007) Electrical signals and their physiological significance in plants. *Plant, Cell & Environment* **30**, 249–257.

Fu, H.Y., Chen, S.J., Chen, R.F., et al. (2006) Identification of oxalic acid and tartaric acid as major persistent pain-inducing toxins in the stinging hairs of the nettle, *Urtica thumbergiana*. *Annals of Botany* **98**, 57–65.

Futuyma, D.J. and Agrawal, A.A. (2009) Macroevolution and the biological diversity of plants and herbivores. *Proceedings of the National Academy of Sciences of the USA* **106**, 18054–18061.

Gershenzon, J. and Dudareva, N. (2007) The function of terpene natural products in the natural world. *Nature Chemical Biology* **3**, 408–414.

Gibson, R.W. and Pickett, J.A. (1983) Wild potato repels aphids by release of aphid alarm pheromone. *Nature* **302**, 608–609.

Gilbert, L. (1975) Ecological consequences of a coevolved mutualism between butterflies and plants. In: Gilbert, L. and Raven, P. (eds), *Coevolution of Animals and Plants*. University of Texas Press, Austin, pp. 210–240.

Gilg, O., Hanski, I. and Sittler, B. (2003) Cyclic dynamics in a simple vertebrate predator-prey community. *Science* **302**, 866–868.

Givnish, T.J. (1990) Leaf mottling: relation to growth form and leaf phenology and possible role as camouflage. *Functional Ecology* **4**, 463–474.

Givnish, T.J., Sytsma, K.J., Smith, J.F. and Hahn, W.J. (1994) Thorn-like prickles and heterophylly in *Cyanea:* adaptations to extinct avian browsers on Hawaii? *Proceedings of the National Academy of Sciences of the USA* **91**, 2810–2814.

Glander, K.E. (1977) Poison in a monkey's garden of Eden. *Natural History* **86**, 35–41.

Glauser, G., Dubugnon, L., Mousavi, S.A.R. et al. (2009) Velocity estimates for signal propagation leading to systemic jasmonic acid accumulation in wounded *Arabidopsis*. *Journal of Biological Chemistry* **284**, 34506–34513.

Glauser, G., Grata, E., Dubugnon, L., et al. (2008) Spatial and temporal dynamics of jasmonate synthesis and accumulation in *Arabidopsis* in response to wounding. *Journal of Biological Chemistry* **283**, 16400–16407.

Godoy-Vitorino, F., Ley, R.E., Gao, Z., et al. (2008) Bacterial community in the crop of the hoatzin, a Neotropical folivorous flying bird. *Applied and Environmental Microbiology* **74**, 5905–5912.

Gonzales-Vigil, E., Bianchetti, C.M., Philipps, G.N. and Howe, G.A. (2011) Adaptive evolution of threonine deaminase in plant defense against insect herbivores. *Proceedings of the National Academy of Sciences of the USA* **108**, 5897–5902.

Grajal, A., Strahl, S.D., Parra, R., et al. (1989) Foregut fermentation in the hoatzin, a Neotropic leaf-eating bird. *Science* **245**, 1236–1238.

Grbić, M., Van Leeuven, T., Clark, R.M., et al. (2011) The genome of *Tetranychus urticae* reveals herbivorous pest adaptations. *Nature* **479**, 487–492.

Green, T.R. and Ryan, C.A. (1972) Wound-induced proteinase inhibitor in plant leaves: a possible defense mechanism against insects. *Science* **175**, 776–777.

Greenwood, R.M. and Atkinson, I.A.E. (1977) Evolution of divaricating plants in New Zealand in relation to moa browsing. *Proceedings of the New Zealand Ecological Society* **24**, 21–33.

Grime, J.P., Cornelissen, J.H.C., Thompson, K. and Hodgson, J.G. (1996) Evidence for a causal connection between anti-herbivore defence and the decomposition rate of leaves. *Oikos* **77**, 489–494.

Grostal, P. and O'Dowd, D.J. (1994) Plants, mites and mutualism: leaf domatia and the abundance and reproduction of mites on *Viburnum tinus* (Caprifoliaceae). *Oecologia* **97**, 308–315.

Grubb, P. (1971) The growth, ecology and population structure of giant tortoises on Aldabra. *Philosophical Transactions of the Royal Society of London, Series B* **260**, 327–372 (see p. 358).

Grubb, P.J. (1992) A positive distrust in simplicity—lessons from plant defences and from competition among plants and among animals. *Journal of Ecology* **80**, 585–610.

Guhling, O., Hobl, B., Yeats, T. and Jetter, R. (2006) Cloning and characterization of a lupeol synthase involved in the synthesis of epicuticular wax crystals on stem and hypocotyl surfaces of *Ricinus communis*. *Archives of Biochemistry and Biophysics* **448**, 60–72.

Gundlach, H., Müller, M.J., Kutchan, T.M. and Zenk, M.H. (1992) Jasmonic acid is a signal transducer in elicitor-induced plant cell cultures. *Proceedings of the National Academy of Sciences of the USA* **89**, 2389–2393.

Haines, W.P. and Renwick, J.A.A. (2009) Bryophytes as food: comparative consumption and utilization of mosses by a generalist insect herbivore. *Entomologia Experimentalis et Applicata* **133**, 296–306.

Hall, D.E., MacGregor, K.B., Nijsse, J. and Bown, A.W. (2004) Footsteps from insect larvae damage leaf surfaces and initiate rapid responses. *European Journal of Plant Pathology* **110**, 441–447.

Halpern, M., Raats, D. and Lev-Yadun, S. (2007) Plant biological warfare: thorns inject pathogenic bacteria into herbivores. *Environmental Microbiology* **9**, 584–592.

Hansen, A.K. and Moran, N.A. (2011) Aphid genome expression reveals host-symbiont cooperation in the production of amino acids. *Proceedings of the National Academy of Sciences of the USA* **108**, 2849–2854.

Hao, B.-Z. and Wu, J.-L. (2000) Laticifer differentiation in *Hevea brasiliensis*: induction by exogenous jasmonic acid and linolenic acid. *Annals of Botany* **85**, 37–43.

Harborne, J.B. (1993) *Introduction to Ecological Biochemistry*, 4th edn. Academic Press, London, p. 113.

Hartl, M., Giri, A.P., Kaur, H.and Baldwin, I.T. (2010) Serine protease inhibitors specifically defend *Solanum nigrum* against generalist herbivores but do not influence plant growth and development. *Plant Cell* 22, 4158–4175.

Hartmann, T. (1999) Chemical ecology of pyrrolizidine alkaloids. *Planta* **207**, 483–495.

Hartmann, T. and Ober, D. (2008) Defense by pyrrolizidine alkaloids: developed by plants and recruited by insects. In: Schaller, A. (ed.), *Induced Plant Resistance to Herbivory*. Springer, Stuttgart, pp. 213–231.

Haslam, E. (1998) *Practical Polyphenolics: From Structure to Molecular Recognition and Physiological Action*. Cambridge University Press, Cambridge.

Healy, J.F. (2004) *Pliny the Elder. Natural History: A Selection*. Penguin Classics, London.

Heil, M. (2009) Damaged self-recognition in plant herbivore defence. *Trends in Plant Science* **14**, 356–363.

Heil, M., Koch, T., Hilpert, A., et al. (2001) Extrafloral nectar production of the ant-associated plant, *Macaranga tanarius*, is an induced, indirect, defensive response elicited by jasmonic acid. *Proceedings of the National Academy of Sciences of the USA* **98**, 1083–1088.

Hejl, A.M., Einhellig, F.A. and Rasmussen, J.A. (1993) Effects of juglone on growth, photosynthesis and respiration. *Journal of Chemical Ecology* **19**, 559–568.

Herms, D.A. and Mattson, W.J. (1992) The dilemma of plants: to grow or defend. *Quarterly Review of Biology* **67**, 283–335.

Hicks, J., Rumpff, H. and Yorkston, H. (1990) *Christmas Crabs*, 2nd edn. Christmas Island Natural History Association, Christmas Island, Australia.

Hilker, M. and Meiners, T. (2010) How do plants 'notice' attack by herbivorous arthropods? *Biological Review of the Cambridge Philosophical Society* 85, 267–280.

Hilker, M., Kobs, C., Varama, M. and Schrank, K. (2002) Insect egg deposition induces *Pinus sylvestris* to attract egg parasitoids. *Journal of Experimental Biology* **205**, 455–461.

Holzer, P. (2011) Transient receptor potential (TRP) channels as drug targets for diseases of the digestive system. *Pharmacology & Therapeutics* **131**, 142–170.

Holzinger, F. and Wink, M. (1996) Mediation of cardiac glycoside insensitivity in the monarch butterfly (*Danaus plexippus*): role of an amino acid substitution in the ouabain binding site of Na^+, K^+-ATPase. *Journal of Chemical Ecology* **22**, 1921–1937.

Horner, H.T., Wanke, S. and Samain, M-S. (2012) A comparison of leaf crystal micropatterns in the two sister genera *Piper* and *Peperomia* (Piperaceae). *American Journal of Botany* **99**, 983–997.

Hou, X., Lee, L.Y.C., Xia, K., et al. (2010) DELLAs modulate jasmonate signaling via competitive binding to JAZs. *Developmental Cell* **19**, 884–894.

Houghton, R.A. (2007) Balancing the global carbon budget. *Annual Review of Earth and Planetary Sciences* **35**, 313–347.

Howe, G.A. and Jander, G. (2008) Plant immunity to insect herbivores. *Annual Review of Plant Biology* **59**, 41–66.

Howe, G.A., Lightner, J., Browse, J. and Ryan, C.A. (1996) An octadecanoid pathway mutant (JL5) of tomato is compromised in signaling for defense against insect attack. *Plant Cell* **11**, 2067–2077.

Huffaker, A., Pearce, G., Veyrat, N., et al. (2013) Plant elicitor peptides are conserved signals regulating direct and indirect antiherbivore defense. *Proceedings of the National Academy of Sciences of the USA* **110**, 5707–5712.

Hughes, S. (1990) Antelope activate the acacia's alarm system. *New Scientist* 29 September, **vol. 127**, p. 19. [based on van Hoven, W. (1984) Trees' secret warning system against browsers. Custos 13, 13–16].

Husebye, H., Chadchawan, S., Winge, P., et al. (2002) Guard cell- and phloem idioblast-specific expression of thioglucoside glucohydrolase 1 (myrosinase) in Arabidopsis. *Plant Physiology* **128**, 1180–1188.

Imhoff, M.L., Bounoua, L., Ricketts, T., et al. (2004) Global patterns in human consumption of net primary production. *Nature* **429**, 870–873.

Jaffré, T., Brooks, R.R., Lee, J. and Reeves, R.D. (1976) *Sebertia acuminata*: a hyperaccumulator of nickel from New Caledonia. *Science* **193**, 579–580.

Jaffré, T., Morat, P., Veillon, J-M., et al. (2001) *Composition and Characterisation of the Native flora of New Caledonia.* Institut de Recherche pour la Développement, Noumea.

Jaffré, T., Pillon, Y., Thomine, S. and Merlot, S. (2013) The metal hyperaccumulators from New Caledonia can broaden our understanding of nickel accumulation in plants. *Frontiers in Plant Science* **4**, Article 279, 1–7.

Janzen, D.H. (1967) Interaction of the bull's-horn acacia (*Acacia cornigera* L.) with an ant inhabitant (*Pseudomyrmex ferruginea* F. Smith) in Eastern Mexico. *University of Kansas Science Bulletin* **XLVII**, 315–558.

Janzen, D.H. (1968) Host plants as islands in evolutionary and contemporary time. *American Naturalist* **102**, 592–595.

Janzen, D.H. (1986) Chihuahuan desert nopaleras: defaunated big mammal vegetation. *Annual Review of Ecology and Systematics* **17**, 595–636.

Janzen, D.H. and Martin, P.S. (1982) Neotropical anachronisms: the fruit the gomphotheres ate. *Science* **215**, 19–27

Johnson, C. (2006) *Australia's Mammal Extinctions: A 50 000 Year History.* Cambridge University Press, Cambridge.

Johnson, C.N. (2009) Ecological consequences of late Quaternary extinctions of megafauna. *Proceedings of the Royal Society B* **276**, 2509–2519.

Johnson, R., Narvaez, J., An, G. and Ryan, C. (1989) Expression of proteinase inhibitors I and II in transgenic tobacco plants: effects on natural defense against *Manduca sexta* larvae. *Proceedings of the National Academy of Sciences of the USA* **86**, 9871–9875.

Jones, K.M.W., Maclagan, S.J. and Krockenberger, A.K. (2006) Diet selection in the green ringtail possum (*Pseudochirops archeri*): a specialist folivore in a diverse forest. *Austral Ecology* **31**, 799–807.

Jongsma, M.A. and Beekwilder, J. (2008) Plant protease inhibitors: functional evolution for defense. In: Schaller, A. (ed.), *Induced Plant Resistance to Herbivory.* Springer, Stuttgart, pp. 235–251.

Jordan, D. (2010) Le Jardin du Talèfre dans le massif de Mont-Blanc à Chamonix. Réévaluation et contribution à sa connaissance botanique. *Le Monde des Plantes* **No. 501**, 9–20.

Juenger, T. and Lennartsson, T. (2000) Tolerance in plant ecology and evolution: toward a more unified theory of plant–herbivore interaction. *Evolutionary Ecology* **14**, 283–287.

Kaiser, W., Huguet, E., Casas, J., et al. (2010) Plant green-island phenotype induced by leaf-miners is mediated by bacterial symbionts. *Proceedings of the Royal Society B* **277**, 2311–2319.

Kalergis, A.M., López, C.B., Becker, M.I., et al. (1997) Modulation of fatty acid oxidation alters contact hypersensitivity to urushiols: role of aliphatic chain beta-oxidation in processing and activation of urushiols. *Journal of Investigative Dermatology* **108**, 57–61.

Kang, K., Pulver, S.R., Panzano, V.C., et al. (2010) Analysis of *Drosophila* TRPA1 reveals an ancient origin for human chemical nociception. *Nature* **464**, 597–600.

Karban, R. and Baldwin, I.T. (1997) *Induced Responses to Herbivory.* University of Chicago Press, Chicago.

Karban, R. and Thaler, J.S. (1999) Plant phase change and resistance to herbivory. *Ecology* **80**, 510–517.

Kelber, A., Vorobyev, M. and Osorio, D. (2003) Animal colour vision—behavioural tests and physiological concepts. *Biological Reviews* **78**, 81–118.

Kessler, A. and Heil, M. (2011) Evolutionary ecology of plant defences: the multiple faces of indirect defences and their agents of natural selection. *Functional Ecology* **25**, 348–357.

Kingdon, J. (1989) *Island Africa. The Evolution of Africa's Rare Animals and Plants.* Princeton University Press, Princeton, NJ, p. 41.

Kingdon, J. (2001) *The Kingdon Field Guide to African Mammals.* Academic Press, San Diego.

Kobayashi, H., Yanaka, M. and Ikeda, T.M. (2010) Exogenous methyl jasmonate alters trichome density on leaf surfaces of Rhodes grass (*Chloris gayana* Kunth). *Journal of Plant Growth Regulation* **29**, 506–511.

Konno, K., Hirayama, C., Yasui, H. and Nakamura, M. (1999) Enzymatic activation of oleuropein: a protein crosslinker used as a chemical defense in the privet tree. *Proceedings of the National Academy of Sciences of the USA* **96**, 9159–9164.

Koo, A.J.K., Gao, X., Jones, A.D. and Howe, G.A. (2009) A rapid wound signal activates the systemic synthesis of bioactive jasmonates in *Arabidopsis. Plant Journal* **59**, 974–986.

Koornneef, A. and Pieterse, C.M. (2008) Cross talk in defense signaling. *Plant Physiology* **146**, 839–844.

Koptur, S. (1992) Extrafloral nectary-mediated interactions between insects and plants. In: Bernays, E.A. (ed.), *Insect-Plant Interactions.* CRC Press, Boca Raton, vol. IV, pp. 81–129.

Koroleva, O.A., Davies, A., Deeken, R., et al. (2000) Identification of a new glucosinolate-rich cell type in *Arabidopsis* flower stalk. *Plant Physiology* **124**, 599–608.

Korth, K.L., Doege, S.J., Park, S.-H., et al. (2006) *Medicago truncatula* mutants demonstrate the role of plant calcium oxalate crystals as an effective defense against chewing insects. *Plant Physiology* **141**, 188–195.

Kotanen, P.M. and Rosenthal, J.P. (2000) Tolerating herbivory: does the plant care if the herbivore has a backbone? *Evolutionary Ecology* **14**, 537–549.

Kowalski, S.P., Eannetta, N.T., Hirzel, A.T. and Steffens, J.C. (1992) Purification and characterization of polyphenol oxidase from glandular trichomes of *Solanum berthaultii. Plant Physiology* **100**, 677–684.

Krimmel, B.A. and Pearse, I.S. (2012) Sticky plant traps insects to enhance indirect defence. *Ecology Letters* **16**, 219–224.

Krumm, T., Bandemer, K. and Boland, W. (1995) Induction of volatile biosynthesis in the lima bean (*Phaseolus lunatus*) by leucine- and isoleucine conjugates of 1-oxo- and 1-hydroxyindan-4-carboxylic acid: evidence for amino acid conjugates of jasmonic acid as intermediates in the octadecanoid signalling pathway. *FEBS Letters* **377**, 523–529.

Kuriyama, A., Kawai, F., Kanamori, M. and Dathe, W. (1993) Inhibitory effect of jasmonic acid on gametophytic growth, initiation and development of sporophytic shoots in *Equisetum arvense. Journal of Plant Physiology* **141**, 694–697.

Labandeira, C.C. (1998) Early history of arthropod and vascular plant associations. *Annual Review of Earth and Planetary Science* **26**, 329–377 (see especially p. 343).

Labandeira, C.C. (2006a) The four phases of plant–arthropod associations in deep time. *Geologica Acta* **4**, 409–438 (see especially p. 417).

Labandeira, C.C. (2006b) Silurian to Triassic plant and hexapod clades and their associations: new data, a review, and interpretations. *Arthropod Systematics and Phylogeny* **64**, 53–94.

Labeyrie, E. and Dobler, S. (2004) Molecular adaptation of *Chrysochus* leaf beetles to toxic compounds in their food plants. *Molecular Biology and Evolution* **21**, 218–221.

Langenheim, J.H. (2003) *Plant Resins: Chemistry Evolution Ecology Ethnobotany.* Timber Press, Portland Oregon.

Lev-Yadun, S. (2001) Aposematic (warning) coloration associated with thorns in higher plants. *Journal of Theoretical Biology* **210**, 385–388.

Lev-Yadun, S., Dafni, A., Flaishman, M.A., et al. (2004) Plant coloration undermines herbivorous insect camouflage. *BioEssays* **26**, 1126–1130.

Levin, D.A. (1976) Alkaloid-bearing plants: an ecogeographic perspective. *American Naturalist* **110**, 262–284.

Liddell Hart, H.B.(1944) *Thoughts on War*. Reissued in 1999 by Spelmount, Staplehurst, Kent, UK.

Little, D., Gouhier-Darimont, C., Bruessow, F. and Reymond, P. (2007) Oviposition by pierid butterflies triggers defense responses in *Arabidopsis*. *Plant Physiology* **143**, 784–800.

Liu, Y., Ahn, J.E., Datta, S., et al. (2005) *Arabidopsis* vegetative storage protein is an anti-insect acid phosphatase. *Plant Physiology* **139**, 1545–1556.

Lopes, G. K., Schulman, H.M. and M. Hermes-Lima (1999) Polyphenol tannic acid inhibits hydroxyl radical formation from Fenton reaction by complexing ferrous ions. *Biochimica and Biophysica Acta* 1472, 142–152.

Lo Piparo, E., Scheib, H., Frei, N., et al. (2008) Flavonoids for controlling starch digestion: structural requirements for inhibiting human α-amylase. *Journal of Medical Chemistry* **51**, 3555–3561.

Losey, J.E. and Vaughan, M. (2006) The economic value of ecological services provided by insects. *Bioscience* **56**, 311–323.

Ma, X., Tan, C., Zhu, D., et al. (2007) Huperzine A from *Huperzia* species—an ethnopharmacological review. *Journal of Ethnopharmacology* **113**, 15–34.

Ma, J.F. and Yamaji, N. (2006) Silicon uptake and accumulation in higher plants. *Trends in Plant Sciences* **11**, 392–397.

MacDougal, J.M. (2003) *Passiflora boenderi* (Passifloraceae), a new egg-mimic passion flower from Costa Rica. *Novon* **13**, 454–458.

Maddox, G.D. and Root, R.B. (1990) Structure of the encounter between goldenrod (*Solidago altissima*) and its diverse insect fauna. *Ecology* **71**, 2115–2124.

Mafli, A., Goudet, J. and Farmer, E.E. (2012) Plants and tortoises: mutations in the *Arabidopsis* jasmonate pathway increase feeding in a vertebrate herbivore. *Molecular Ecology* **21**, 2534–2541.

Mäntylä, E., Alessio, G.A., Blande, J.D., et al. (2008) From plants to birds: higher avian predation rates in trees responding to insect herbivory. *PLoS One* e2832.

Marsh, K.J., Wallis, I.R. and Foley, W.J. (2005) Detoxification rates constrain feeding in common brushtail possums (*Trichosurus vulpecula). Ecology* **86**, 2946–2954.

Massey, A.B. (1925) Antagonism of the walnuts (*Juglans nigra* L. and *J. cinerea* L.) in certain plant associations. *Phytopathology* **15**, 773–784.

Massey, F.P. and Hartley, S.E. (2009) Physical defences wear you down: progressive and irreversible impacts of silica on insect herbivores. *Journal of Animal Ecology* **78**, 281–291.

Massey, F.P., Ennos, A.R. and Hartley, S.E. (2007) Herbivore specific induction of silica-based plant defences. *Oecologia* **152**, 677–683.

Massey, F.P., Smith, M.J., Lambin, X. and Hartley, S.E. (2008) Are silica defences in grasses driving vole population cycles? *Biology Letters* **4**, 419–422.

Mauricio, R. and Rauscher, M.D. (1997) Experimental manipulation of putative selective agents provides evidence for the role of natural enemies in the evolution of plant defense. *Evolution (Lawrence, Kansas)* **51**, 1435–1444.

McConn, M., Creelman, R.A., Bell, E., et al. (1997) Jasmonate is essential for insect defense in *Arabidopsis. Proceedings of the National Academy of Sciences USA* **94**, 5473–5477.

McGlone, M.S. and Clarkson, B.D. (1993) Ghost stories: moa, plant defences and evolution in New Zealand. *Tuatara* **32**, 1–21.

McGurl, B., Pearce, G., Orozco-Cardenas, M. and Ryan, C.A. (1992) Structure, expression, and antisense inhibition of the systemin precursor gene. *Science* **255**, 1570–1573.

McNair, M.R. (2003) The hyperaccumulation of metals by plants. *Advances in Botanical Research* **40**, 63–105.

McNaughton, S.J. and Tarrants, J.L. (1983) Grass leaf silification: natural selection for an inducible defense against herbivores. *Proceedings of the National Academy of Sciences of the USA* **80**, 790–791.

McQueen, D.R. (2000) Divaricating shrubs in Patagonia and New Zealand. *New Zealand Journal of Ecology* **24**, 69–80.

Meehan, C.J., Olson, E.J., Reudink, M.W., et al. (2009) Herbivory in a spider through exploitation of an ant-plant mutualism. *Current Biology* **19**, R892–R893.

Mehansho, H., Butler, L.G. and Carlson, D.M. (1987) Dietary tannins and salivary proline-rich proteins: interactions, induction, and defense mechanisms. *Annual Review of Nutrition* **7**, 423–440.

Meyer, G.A. and Root, R.B. (1993) Effects of herbivorous insects and soil fertility on reproduction of goldenrod. *Ecology* **74**, 1117–1128.

Milewski, A.V., Young, T.P. and Madden, D. (1991) Thorns as induced defenses: experimental evidence. *Oecologia* **86**, 70–75.

Miller, A. and Morris, M. (2004) *Ethnoflora of the Soqotra Archipelago*. Royal Botanic Garden Edinburgh, Edinburgh.

Milton, K. (1979) Factors influencing leaf choice by howler monkeys: a test of some hypotheses of food selection by generalist herbivores. *American Naturalist* **114**, 362–378.

Mineur, Y.S., Abizaid, A., Rao, Y., et al. (2011) Nicotine decreases food intake through activation of POMC neurons. *Science* **332**, 1330–1332.

Mithöfer, A., Wanner, G. and Boland, W. (2005) Effects of feeding *Spodoptera littoralis* on Lima bean leaves. II. Continuous mechanical wounding resembling insect feeding is sufficient to elicit herbivory-related volatile emission. *Plant Physiology* **137**, 1160–1168.

Mitter, C., Farrell, B. and Wiegmann, B. (1988) The phylogenetic study of adaptive zones: has phytophagy promoted insect diversification? *American Naturalist* **132**, 107–128.

Moles, A.T., Bonser, S.P, Poore, A.G.B., et al. (2011) Evolutionary ecology of plant defences: assessing the evidence for latitudinal gradients in plant defence and herbivory. *Functional Ecology* **25**, 380–388.

Moore, B.D., Lawler, I.R., Wallis, I.R., et al. (2010) Palatability mapping: a koala's eye view of spatial variation in habitat quality. *Ecology* **91**, 3165–3176.

Moreno, J.E., Tao, Y., Chory, J. and Ballaré, C.L. (2009) Ecological modulation of plant defense via phytochrome control of jasmonate sensitivity. *Proceedings of the National Academy of Sciences of the USA* **106**, 4935–4940.

Mousavi, A.R., Chauvin, A., Pascaud, F., et al. (2013) *GLUTAMATE RECEPTOR-LIKE* genes mediate leaf-to-leaf wound signalling. *Nature*, **500**, 422–426.

Müller, R., de Vos, M., Sun, J.Y., et al. (2010) Differential effects of indole and aliphatic glucosinolates on lepidopteran herbivores. *Journal of Chemical Ecology* **36**, 905–913.

Mumm, R. and Dicke, M. (2010) Variation in natural plant products and the attraction of bodyguards involved in indirect plant defense. *Canadian Journal of Zoology* **88**, 628–667.

Murphy, R.J. and Alvin, K.L. (1997) Fibre maturation in the bamboo *Giganochloa scortechinii*. *IAWA Journal* **18**, 147–156.

Murton, R.K., Isaacson, A.J. and Westwood, N.J. (1966) The relationships between wood-pigeons and their clover food supply and the mechanisms of population control. *Journal of Applied Ecology* **3**, 55–96.

Nabhan, G.P. (1997) Why chilies are hot. *Natural History* **106**, 24–29.

Nakata, P.A. (2012) Engineering calcium oxalate crystal formation in Arabidopsis. *Plant and Cell Physiology* **53**, 1275–1282.

Naumkin, V.V. (1993) *Island of the Phoenix. An Ethnographic Study of the People of Socotra*. Ithaca Press, Reading, MA.

Nelson, S.L., Kunz, T.H. and Humphrey, S.R. (2005) Folivory in fruit bats: leaves provide a natural source of calcium. *Journal of Ecology* **31**, 1638–1691.

Nishida, R. (2002) Sequestration of defense substances from plants by lepidoptera. *Annual Review of Entomology* **47**, 57–92.

Nitta, I., Kida, A., Fujibayashi, Y., et al. (2006) Calcium carbonate deposition in a cell wall sac formed in mulberry idioblasts. *Protoplasma* **228**, 201–208.

Novotny, V., Basset, Y., Miller, S.E., et al. (2002) Low host specificity of herbivorous insects in a tropical forest. *Nature* **416**, 841–844.

Osnas, J.L.D., Lichstein, J.W., Reich, P.B. and Pacala, S.W. (2013) Global leaf trait relationships: mass, area, and the leaf economics spectrum. *Science* **340**, 741–744.

Owen-Smith, N. and Novellie, P. (1982) What should a clever ungulate eat? *American Naturalist* **119**, 151–178.

Peiffer, M., Tooker, J.F., Luthe, D.S. and Felton, G.W. (2009) Plants on early alert: glandular trichomes as sensors for insect herbivores. *New Phytologist* **184**, 644–656.

Persson, I.-L., Danell, K. and Bergström, R. (2000) Disturbance by large herbivores in boreal forests with special reference to moose. *Annales Zoologici Fennici* **37**, 251–263.

Phillips, S.J. and Comus, P.W. (eds) (2000) *A Natural History of the Sonoran Desert*. Arizona-Sonora Desert Museum Press, Tucson, p. 136.

Pickard, W.F. (2008) Laticifers and secretory ducts: two other tube systems in plants. *New Phytologist* **177**, 877–888.

Piperno, D.R. (2006) *Phytoliths: A Comprehensive Guide for Archeologists and Paleoecologists*. AltaMira Press, New York.

Pollard, A.J. and Briggs, D. (1984a) Genecological studies of *Urtica dioica* L. II. Patterns of variation at Wicken Fen. *New Phytologist* **96**, 483–499.

Pollard, A.J. and Briggs, D. (1984b) Genecological studies of *Urtica dioica* L. III. Stinging hairs and plant-herbivore interactions. *New Phytologist* **97**, 507–522.

Ponce de León, I., Schmelz, E.A., Gaggero, C. et al. (2012) *Physcomitrella patens* activates reinforcement of the cell wall, programmed cell death and accumulation of evolutionary conserved defence signals, such as salicylic acid and 12-oxo-phytodienoic acid, but not jasmonic acid, upon *Botrytis cinerea* infection. *Molecular Plant Pathology* 13, 960–974.

Price, P.W., Bouton, C.E., Gross, P., et al. (1980) Interactions among three trophic levels: influence of plants on interactions between insect herbivores and natural enemies. *Annual Review of Ecology and Systematics* **11**, 41–65.

Prins, H.H.T. and Gordon, I.J. (2008) Introduction: grazers and browsers in a changing world. In: Gordon, I.J. and Prins, H.H.T. (eds), *The Ecology of Browsing and Grazing*. Springer, Berlin, pp. 1–20.

Prychid, C.J. and Rudall, P.J. (1999) Calcium oxalate crystals in monocotyledons: a review of their structure and systematics. *Annals of Botany* **84**, 725–739.

Pullin, A.S. and Gilbert, J.E. (1989) The stinging nettle, *Urtica dioica*, increases trichome density after herbivore and mechanical damage. *Oikos* **54**, 275–280.

Quinn, C.F., Freeman, J.L., Galeas, M.L., et al. (2008) The role of selenium in protecting plants against prairie dog herbivory: implications for the evolution of selenium hyperaccumulation. *Oecologia* **155**, 267–275.

Racine, C.H. and Downhower, J.F. (1974) Vegetative and reproductive strategies of *Opuntia* (Cactaceae) in the Galapagos Islands. *Biotropica* **6**, 175–186.

Rascio, N. and Navari-Izzo, F. (2011) Heavy metal hyperaccumulating plants: how and why do they do it? And what makes them so interesting? *Plant Science* **180**, 169–181.

Rask, L., Andréasson, E., Ekbom, B., et al. (2000) Myrosinase: gene family evolution and herbivore defense in Brassicaceae. *Plant Molecular Biology* **42**, 93–113.

Rasmann, S. and Agrawal, A.A. (2011) Latitudinal patterns in plant defense: evolution of cardenolides, their toxicity and induction following herbivory. *Ecology Letters* **14**, 476–483.

Rauh, W. (1943) Beiträge zur Morphologie und Biologie der Holzgewächse. II. Morphologische Beobachtungen an Dorngehölzen. *Botanisches Archiv* **43**, 111–168.

Redfern, M. (2011) *Collins New Naturalist Library (117)—Plant Galls*. Collins, London.

Reeds, P.J. (2000) Dispensable and indispensable amino acids for humans. *Journal of Nutrition* **130**, 1835S–1840S.

Reinbothe, S., Reinbothe, C., Lehmann, J., et al. (1994) JIP60, a methyl jasmonate-induced ribosome-inactivating protein involved in plant stress reactions. *Proceedings of the National Academy of Sciences of the USA* **91**, 7012–7016.

Rempt, M. and Pohnert, G. (2010) Novel acetylenic oxylipins from the moss *Dicranum scoparium* with antifeeding activity against herbivorous slugs. *Angewandte Chemie* (international edition in English) **49**, 4755–4758.

Renwick, J.A.A. (2002) The chemical world of crucivores: lures, treats, and traps. *Entomologia Experimentalis et Applicata* **104**, 35–42.

Reymond, P., Bodenhausen, N., Van Poecke, R.M.P., et al. (2004) A conserved transcript pattern in response to a specialist and a generalist herbivore. *Plant Cell* **16**, 3132–3147.

Reymond, P., Weber, H., Damond, M. and Farmer, E.E. (2000) Differential gene expression in response to mechanical wounding and insect feeding in *Arabidopsis. Plant Cell* **12**, 707–719.

Rhoades, D.F. and Cates, R.G. (1976) Toward a general theory of plant antiherbivory chemistry. In *Recent Advances in Phytochemistry*. In: Wallace J.W. and Mansell, R.L. (eds), Plenum Press, New York, **vol. 10**, pp. 168–213.

Rico-Gray, V. and Oliveira, P.S. (2007) *The Ecology and Evolution of Ant-Plant interaction*. University of Chicago Press, Chicago, pp. 99–141.

Riddle, J.M. (1997) *Eve's Herbs: A History of Contraception and Abortion in the West*. Harvard University Press, Cambridge, MA [and Bilger, B. (1998) The secret garden. *The Sciences* January/February 38–43].

Romero, G.Q. and Benson, W.W. (2005) Biotic interactions of mites, plants and leaf domatia. *Current Opinion in Plant Biology* **8**, 436–440.

Romero, G.Q., Souza, J.C. and Vasconcellos-Neto, J. (2008) Antiherbivore protection by mutualistic spiders and the role of plant glandular trichomes. *Ecology* **89**, 3105–3115.

Root, R.B. and Cappuccino, N. (1992) Patterns in population change and the organization of the insect community associated with goldenrod. *Ecological Monographs* **62**, 393–420.

Ryan, C.A. (1974) Assay and biochemical properties of the proteinase inhibitor inducing factor, a wound hormone. *Plant Physiology* **54**, 328–332.

Ryan, C. A. (1990) Proteinase inhibitors in plants: genes for improving defenses against insects and pathogens. *Annual Review of Phytopathology* **28**, 425–449.

Ryan, C.A. and Farmer, E.E. (1999) Method of inducing plant defense mechanisms. US Patent Serial No. 5,935,809 filed 25 July 1994.

Saatchi, S.S., Harris, N.L., Brown, S., et al. (2011) Benchmark map of forest carbon in tropical regions across three continents. *Proceedings of the National Academy of Sciences of the USA* **108**, 9899–9904.

Sagner, S., Kneer, R., Wanner, G., et al. (1998) Hyperaccumulation, complexation and distribution of nickel in *Sebertia acuminata. Phytochemistry* **47**, 339–347.

Saniewski, M., Miyamoto, K. and Ueda, J. (1998) Methyl jasmonate induces gums and stimulates anthocyanin accumulation in peach shoots. *Journal of Plant Growth Regulation* **17**, 121–124.

Schemske, D.W., Mittelbach, G.G., Cornell, H.V., et al. (2009) Is there a latitude gradient in the importance of biotic interactions? *Annual Review of Ecology, Evolution and Systematics* **40**, 245–269.

Schildknecht, H. and Rauch, G. (1961) Die chemische Natur der «Luftphytoncide» von Blattpflanzen insbesondere von *Robinia pseudacacia. Zeitschrift für Naturforschung* **16b**, 422–429.

Schlüter, U., Benchabane, M., Munger, A., et al. (2010) Recombinant proteinase inhibitors for pest control: a multitrophic perspective. *Journal of Experimental Botany* **61**, 4169–4189.

Schmitt, B., Schultz, H., Storsberg, J. and Keusgen, M. (2005) Chemical characterization of *Allium ursinum* L. depending on harvesting time. *Journal of Agricultural and Food Chemistry* **53**, 7288–7294.

Schneider, H., Schuettpelz, E., Pryer, K.M., et al. (2004) Ferns diversified in the shadow of angiosperms. *Nature* **428**, 553–557.

Schofield, R.M., Nesson, M.H. and Richardson, K.A. (2002) Tooth hardness increases with zinc-content in mandibles of young adult leaf-cutter ants. *Naturwissenschaften* **89**, 579–583.

Schoonhoven, L.M., van Loon, J.J.A. and Dicke, M. (2005) *Insect-Plant Biology*, 2nd edn. Oxford University Press, Oxford.

Schuler, M.A. (1996) The role of cytochrome P450 monooxygenases in plant-insect interactions. *Plant Physiology* **112**, 1411–1419.

Sembdner, G. and Parthier, B. (1993) The biochemistry and the physiological and molecular actions of jasmonates. *Annual Review of Plant Physiology and Plant Molecular Biology* 44, 569–589.

Senebier, J. (1800) *Physiologie Végétale*. J.J. Paschoud, Genève, vol. I, pp. 374–375 [my translation. Senebier was referring to 'chardon', any of three genera: *Cirsium, Carlina or Cardus*].

Sévenet, T. and Pusset, J. (1996) Alkaloids from the medicinal plants of New Caledonia. *The Alkaloids: Chemistry and Pharmacology* **48**, 1–73.

Shaw, M.W. and Haughs, G.M. (1983) Damage to potato foliage by *Sminthurus viridis* (L.). *Plant Pathology* **32**, 465–466.

Sheard, L.B., Tan, X., Mao, H., et al. (2010) Jasmonate perception by inositol-phosphate potentiated COI1-JAZ co-receptor. *Nature* **468**, 400–405.

Shroff, R., Vergara, F., Muck, A., et al. (2008) Nonuniform distribution of glucosinolates in *Arabidopsis thaliana* leaves has important consequences for plant defense. *Proceedings of the National Academy of Sciences of the USA* **105**, 6196–6201.

Signer, C., Ruf, T. and Arnold, W. (2011) Hypometabolism and basking: the strategies of Alpine ibex to endure harsh over-wintering conditions. *Functional Ecology* **25**, 537–547.

Smith, A.P. (1986) Ecology of leaf color polymorphism in a tropical forest species: habitat segregation and herbivory. *Oecologia* **69**, 283–287.

Soltau, U., Dötterl, S. and Liede-Schumann, S. (2009) Leaf variegation in *Caladium steudneriifolium* (Araceae): a case of mimicry? *Evolutionary Ecology* **23**, 503–512.

Stahl, E. (1888) *Pflanzen und Schnecken: Biologische Studie über die Schutzmittel der Pflanzen gegen Schneckenfrass*. Gustav Fischer, Jena.

Staswick, P.E. and Tiryaki, I. (2004) The oxylipin signal jasmonic acid is activated by an enzyme that conjugates it to isoleucine in *Arabidopsis*. *Plant Cell* **16**, 2117–2127.

Steppuhn, A. and Baldwin, I.T. (2007) Resistance management in a native plant: nicotine prevents herbivores from compensating for plant protease inhibitors. *Ecology Letters* **10**, 499–511.

Steppuhn, A., Gase, K., Krock, B., et al. (2004) Nicotine's defensive function in nature. *PLoS Biology* **2**, e217.

Strauss, S.Y. and Agrawal, A.A. (1999) The ecology and evolution of plant tolerance to herbivory. *Trends in Ecology and Evolution* **14**, 179–185.

Stumpe, M., Göbel, C., Faltin, B., et al. (2010) The moss *Physcomitrella patens* contains cyclopentenones but no jasmonates: mutations in allene oxide cyclase lead to reduced fertility and altered sporophyte morphology. *New Phytologist* **188**, 740–749.

Szallasi, A. and Blumberg, P.M. (1989) Resiniferatoxin, a phorbol-related diterpene, acts as an ultrapotent analog of capsaicin, the irritant constituent in red pepper. *Neuroscience* **30**, 515–520.

Tennie, C., Hedwig, D., Call, J.and Tomasello, M. (2008) An experimental study of nettle feeding in captive gorillas. *American Journal of Primatology* **70**, 1–10.

Terashima, I., Fujita, T., Inoue, T., et al. (2009) Green light drives leaf photosynthesis more efficiently than red light in strong white light: revisiting the enigmatic question of why leaves are green. *Plant & Cell Physiology* **50**, 684–687.

Terborgh, J., Lopez, L., Nuñez, V., et al. (2001) Ecological meltdown in predator-free forest fragments. *Science* **294**, 1923–1926.

Tewksbury, J.J. and Nabhan, G.P. (2001) Seed dispersal. Directed deterrence by capsaicin in chillies. *Nature* **412**, 403–404.

Textor, S. and Gershenzon, J. (2009) Herbivore induction of the glucosinolate—myrosinase defense system: major trends, biochemical bases and ecological significance. *Phytochemical Reviews* **8**, 149–170.

Thaler, J.S. (1999) Induced resistance in agricultural crops: effects of jasmonic acid on herbivory and yield in tomato plants. *Environmental Entomology* **28**, 30–37.

Thayer, S.S. and Conn, E.E. (1981) Subcellular localization of dhurrin β-glucosidase and hydroxynitrile lyase in mesophyll cells of Sorghum leaf blades. *Plant Physiology* **67**, 617–622.

Theodossiou, T.A., Hothersall, J.S., De Witte, P.A., et al. (2009) The multifaceted photocytotoxic profile of hypericin. *Molecular Pharmacology* **6**, 1775–1789.

Thines, B., Katsir, L., Melotto, M., et al. (2007) JAZ repressor proteins are targets of the SCF(COI1) complex during jasmonate signalling. *Nature* **448**, 661–665.

Thulin, M. and Warfa, A.M. (1987) The frankincense trees (*Boswellia* spp., *Burseraceae*) of northern Somalia and southern Arabia. *Kew Bulletin* **42**, 487–500.

Thurston, E.L. (1974) Morphology, fine structure, and ontogeny of the stinging emergence of *Urtica dioica*. *American Journal of Botany* **61**, 809–817.

Thurston, E.L. (1976) Morphology, fine structure and ontogeny of the stinging emergence of *Tragia ramosa* and *T. Saxicola* (Euphorbiaceae). *American Journal of Botany* **63**, 710–718.

Thurston, E.L. and Lersten, N.R. (1969) The morphology and toxicology of plant stinging hairs. *Botanical Review* **35**, 393–412.

Timm, R.M. and Vriesendorp, C. (2003) Observations on feeding behaviour in the vesper mouse, *Nyctomys sumichrasti*. *Mammalian Biology* **68**, 126–128.

Tokunaga, T., Takada, N. and Ueda, M. (2004) Mechanism of antifeedant activity of plumbagin, a compound concerning the chemical defense in carnivorous plants. *Tetrahedron Letters* **45**, 7115–7119.

Trapp, S. and Croteau, R. (2001) Defensive resin biosynthesis in conifers. *Annual Review of Plant Physiology and Plant Molecular Biology* **52**, 689–724.

Traw, M.B. and Bergelson, J. (2003) Interactive effects of jasmonic acid, salicylic acid, and gibberellin on induction of trichomes in *Arabidopsis*. *Plant Physiology* **133**, 1367–1375.

Tuberville, T.D., Dudley, P.G. and Pollard, A.J. (1996) Responses of invertebrate herbivores to stinging trichomes of *Urtica dioica* and *Laportea canadensis*. *Oikos* **75**, 83–88.

Turlings, T.C.J., Lengwiler, U.B., Bernasconi, M.L. and Wechsler, D. (1998) Timing of induced volatile emissions in maize seedlings. *Planta* **207**, 146–152.

Turlings, T.C.J., Tumlinson, J. H., Heath, R.R., et al. (1991) Isolation and identification of allelochemicals that attract the larval parasitoid, *Cotesia marginiventris* (Cresson), to the microhabitat of one of its hosts. *Journal of Chemical Ecology* **17**, 2235–2251.

Turlings, T.C.J., Tumlinson, J.H. and Lewis, W.J. (1990) Exploitation of herbivore-induced plant odors by host-seeking parasitic wasps. *Science* **250**, 1251–1253.

Tzu, S. (1988) *The Art of War*. Shambhala Publications, Boston (translated by T. Cleary).

Ueda, J. and Kato, J. (1980) Isolation and identification of a senescence-promoting substance from Wormwood (*Artemisia absinthium* L.). *Plant Physiology* **66**, 246–249.

Ueda, H., Nishiyama, C., Shimada, T., et al. (2006) AtVAM3 is required for normal specification of idioblasts, myrosin cells. *Plant and Cell Physiology* **47**, 164–175.

Van Bel, A.J.E. (2003) The phloem, a miracle of ingenuity. *Plant, Cell & Environment* **26**, 125–149 (see especially pp. 131–132).

Van Damme, K. and Banfield, L. (2011) Past and present human impacts on biodiversity of Socotra Island (Yemen): implications for future conservation. In: Knight M., Mallon D., Seddon P. (eds), Biodiversity Conservation in the Arabian Peninsula. *Zoology in the Middle East Supplement* **3**, pp. 31–88.

Van der Meijden, E., Wijn, M. and Verkaar, H.J. (1988) Defense and regrowth, alternative strategies in the struggle against herbivores. *Oikos* **51**, 355–363.

VanEtten, H.D., Mansfield, J.D., Bailey, J.A. and Farmer, E.E. (1994) Two classes of plant antibiotics: phytoalexins and phytoanticipins. *Plant Cell* 6, 1191–1192.

van Hoven, W. (1991) Mortalities in kudu (*Tragelaphus strepsiceros*) populations related to chemical defence in trees. *Journal of African Zoology* **105**, 141–145.

Van Marken Lichtenbelt, W.D. (1993) Optimal foraging of a herbivorous lizard, green iguana, in a seasonal environment. *Oecologia* 95, 246–256.
Van Schaik, C.P. and Griffiths, M. (1996) Activity periods of Indonesian rain forest mammals. *Biotropica* **28**, 105–112.
Van Soest P.J. (1996) Allometry and ecology of feeding behavior and digestive capacity in herbivores: a review. *Zoo Biology* **15**, 455–479.
Vannini, M. and Ruwa, R.K. (1994) Vertical migrations in the tree crab *Sesarma leptosoma* (Decapoda, Grapsidae). *Marine Biology* **118**, 271–278.
Vetter, J. (2000) Plant cyanogenic glycosides. *Toxicon* **38**, 11–36.
Vick, B.A. and Zimmerman, D.C. (1984) Biosynthesis of jasmonic acid by several plant species. *Plant Physiology* **75**, 458–461.
Vrieling, K., Smit, W. and van der Meijden, E. (1991) Tritrophic interactions between aphids (*Aphis jacobaeae* Schrank), ant species, *Tyria jacobaeae* L., and *Senecio jacobaea* L. lead to maintenance of genetic variation in pyrrolizidine alkaloid concentration. *Oecologia* **86**, 177–182.
Vriens, J., Nilius, B. and Vennekens, R. (2008) Herbal compounds and toxins modulating TRP channels. *Current Neuropharmacology* **6**, 79–96.
Wagner, G.J. (1991) Secreting glandular trichomes: more than just hairs. *Plant Physiology* **96**, 675–679.
Waldhoer, M., Bartlett, S.E. and Whistler, J.L. (2004). Opioid receptors. *Annual Review of Biochemistry* **73**, 953–990.
Walter, D.E. (1996) Living on leaves: mites, tomenta and leaf domatia. *Annual Review of Entomology* **41**, 101–114.
Wasternack, C. and Hause, B. (2013) Jasmonates: biosynthesis, perception, signal transduction and action in plant stress response, growth and development. *Annals of Botany* **111**, 1021–1058.
Welp, L.R., Keeling, R.F., Meijer, H.A.J., et al. (2011) Interannual variability in the oxygen isotopes of atmospheric CO2 driven by El Niño. *Nature* **477**, 579–582.
Wess, J., Eglen, R.M. and Gautam, D. (2007) Muscarinic acetylcholine receptors: mutant mice provide new insights for drug development. *Nature Reviews Drug Discovery* **6**, 721–733.
Wheat, C.W., Vogel, H., Wittstock, U., et al. (2007) The genetic basis of plant-insect coevolutionary key innovation. *Proceedings of the National Academy of Sciences of the USA* **104**, 20427–20431.
Wheeler, K.G.R. (2005) *A Natural History of Nettles*. Trafford Publishing, Victoria, Canada.
Wiens, D. (1978) Mimicry in plants. *Evolutionary Biology* **11**, 365–403.
Wildon, D.C., Thain, J.F., Minchin, P.E.H., et al. (1992) Electrical signalling and systemic proteinase inhibitor induction in the wounded plant. *Nature* **360**, 62–65.
Will, T., Tjallingi, W.F., Thönnessen, A. and van Bel, A.J.E. (2007) Molecular sabotage of plant defense by aphid saliva. *Proceedings of the National Academy of Sciences of the USA* **104**, 10536–10541.
Wilson, C.R. and Hooser, S.B. (2007) Toxicity of yew (*Taxus* spp.) alkaloids. In: Gupta, R.C. (ed.), *Veterinary Toxicology. Basic and Clinical Principles*. New York: Academic Press; pp. 929–935.
Wink, M. (2008) Plant secondary metabolism: diversity, function and its evolution. *Natural Product Communications* **3**, 1205–1216.
Wittstock, U., Agerbirk, N., Stauber, E.J., et al. (2004) Successful herbivore attack due to metabolic diversion of a plant chemical defense. *Proceedings of the National Academy of Sciences of the USA* **101**, 4859–4864.
Wittstock, U. and Burow, M. (2010) Glucosinolate breakdown in Arabidopsis: mechanism, regulation and biological significant. *The Arabidopsis Book* **8**, 1–14.
Worthy, T.H. and Holdaway, R.N. (2002) *The Lost World of the Moa*. Indiana University Press, Bloomington, IN.
Wright, I.J., Reich, P.B., Westoby, M., et al. (2004) The worldwide leaf economics spectrum. *Nature* **428**, 821–827.
Wu, C-C., Chen S.-J., Yen T.-B. and Kuo-Huang, L.-L. (2006) Influence of calcium availability on deposition of calcium carbonate and calcium oxalate crystals in the idioblasts of *Morus australis* Poir. leaves. *Botanical studies* **47**, 119–127.

Xie, D-X., Feys, B.F., James, S., et al. (1998) *COI1*: an *Arabidopsis* gene required for jasmonate-regulated defense and fertility. *Science* **280**, 1091–1094.

Xiu, X., Puskar, N.L.Shanata, J.A.P., et al. (2009) Nicotine binding to brain receptors requires a strong cation-interaction. *Nature* **458**, 534–537.

Yamada, K., Ojika, M. and Kigoshi, H. (2007) Ptaquiloside, the major toxin of bracken, and related terpene glycosides: chemistry, biology and ecology. *Natural Product Reports* **24**, 798–813.

Yan, Q. and Bennick, A. (1995) Identification of histatins as tannin-binding proteins in human saliva. *Biochemical Journal* **311**, 341–347.

Yan, Y., Stolz, S., Chételat, A., et al. (2007) A downstream mediator in the growth repression limb of the jasmonate pathway. *Plant Cell* **19**, 2470–2483.

Yang, D.-L., Yao, J., Mei, C.-S., et al. (2012) Plant hormone jasmonate prioritizes defense over growth by interfering with gibberellin signaling cascade. *Proceedings of the National Academy of Sciences of the USA* **109**, E1192–E1200

Yukimune, Y., Tabata, H., Higashi, Y. and Hara, Y. (1996) Methyl jasmonate-induced overproduction of paclitaxel and baccatin III in *Taxus* cell suspension cultures. *Nature Biotechnology* **14**, 1129–1132.

Zagrobelny, M., Bak, S. and Møller, L. (2008) Cyanogenesis in plants and arthropods. *Phytochemistry* **69**, 1457–1468.

Zarate, S.I., Kempema, L.A. and Walling, L.L. (2007) Silverleaf whitefly induces salicylic acid defenses and suppresses effectual jasmonic acid defenses. *Plant Physiology* **143**, 866–875.

Zavala, J.A., Patankar, A.G., Gase, K., Baldwin, I.T. (2004). Constitutive and inducible trypsin proteinase inhibitor production incurs large fitness costs in *Nicotiana attenuata. Proceedings of the National Academy of Sciences USA* **101**, 1607–1612.

Zhao, Z., Zhang, W., Stanley, B.A. and Assmann, S.M. (2008) Functional proteomics of *Arabidopsis thaliana* guard cells uncovers new stomatal signaling pathways. *Plant Cell* **20**, 3210–3226.

Zimmerman, L.C. and Tracy, C.R. (1989) Interactions between the environment and ectothermy and herbivory in reptiles. *Physiological Zoology* **62**, 374–409.

Index

Page numbers in *italics* indicate illustrations.

A
abortifacients 111
Abramites spp. (headstander fish), herbivory 12
Acacia peuce, (Fabaceae; waddywood), anachronistic defences 160
acacias 120, 154, 160, 170
see also Vachellia spp.
Acanthaceae, on Socotra 170–1
Acantholimon spp. (Plumbaginaceae), in thorn cushion 59
Acanthophyllum spp. (Caryophyllaceae), in thorn cushion 59
Acanthospermum hispidum, seed dispersal by mammals 172
acetylcholine receptor antagonists 89
Aciphylla spp. (Apiaceae; New Zealand spear grasses) 41–2, 158–9, 173
Adenium obesum (Apocynaceae; desert rose) *167* (*Plate 13*)
Agave spp. (Agavaceae), fibre cells 68
agriculture, relevance of plant defences 180–1
Aizoaceae, stone mimicry 41
Alces alces (moose), collateral damage to plants 17
Aldabra (Seychelles), tortoise food plants 174
Aleurites rockinghamensis (Euphorbiaceae), latex production 72
alien herbivores, non-adapted plants 24
alkaloids
from evergreens 89
from Papaveraceae 88
from Solanaceae 88–9
geographical distribution 90, *91*
targets 89
toxicity 86–8
Allium ursinum (Liliaceae; wild bear garlic), defence-related odour 79
Alouatta spp. (howler monkeys)
alkaloid poisoning 87–8
choice of leaves 22–3
Alseuosmia pusilla (Alseuosmiaceae), mimicry 42
amber 105–6
Amborella trichopoda (Amborellaceae), methyl salicylate 90
amino acids
targetted destruction 116–17
valued resource for humans 181–2
amphibians, lack of folivory 11
Amyema cambagei (mistletoe), host mimicry 42, *43*
amylase inhibitors 118–19
anachronistic defences 155, 159–61
Andira inermis (Fabaceae), andirine alkaloid 87
Angkalanthus oligophylla (Acanthaceae), heterophylly on Socotra 171–2
Anisotes diversifolius (Acanthaceae), polymorphism on Socotra 171
Annona muricata (Annonaceae, custard apple), domatia 149
ant—plant interactions 143–7
aphids 7, 21, 115, 142–3
aposematic patterning 33
apparency theory 161–2
Arabidopsis
engineered calcium oxalate synthesis 67
engineered monoterpene synthesis 104
fou2 wound-mimic mutant 121–2
glucosinolates 85
inducible defence chemicals (arabidopsides) *119*
insect-induced gene expression 11
jasmonate mutant studies 124–6
myrosin cells 70
non-standard amino acids 116
response to injury 130–2
stomatal guard cells 48
trichomes 48, 50–1
Araucaria spa., selection for spininess 60–1
Arborimus longicaudus (red tree vole) 106, 162
Arecaceae (rattans) 143–4
Argemone mexicana (Mexican prickly poppy), toxic alkaloids 88
arginine, targeted by plant proteins 115
aristolochic acid *82*
Artemisia spp. (Asteraceae; wormwoods), trichomes 48
Arytera divaricata (Sapindaceae), food for exudivores 72
Asclepias spp. (milkweeds), specialist feeders 109
Astragalus spp. (Fabaceae) 59, 79, 87
Atkinson, I.A.E. 156–7
Atropa belladonna (Solanaceae; deadly nightshade), scopolamine 89
Australia
Araucaria spp. 60
folivorous marsupials 16, 61, 64, 72, 101, 162
human influences on fauna 160
Loranthaceae (mistletoes) 42–3
stinging plants 52–4
axillary tuft domatia 148

B
Bacillus cereus, found on date palm 62
Bagheera kiplingi (spider), herbivorous 146
Baldwin, Ian 90
Balfour, Isaac Bayley 170
Barleria spp. (Acanthaceae), thorns on Socotra 168, 170, *171* (*Plate 15*)
barley, inducible ribosome inactivating proteins (RIPs) 116
Barlow, Bryan 42–3
Barteria fistulosa (Passifloraceae), ant plant 147
bats, feeding from leaves 16
Beltian bodies *145*, 146
Benrey, Betty 151
Betula pubescens (Betulaceae; mountain birch), volatiles 153
birds
  attracted by volatiles 153
  colour vision 31
  and divarication 156–7
  flightlessness 13
  folivorous 13, 173
  legendary 173
  visually guided when feeding 158
Blepharis ciliaris (Acanthaceae), thorns on Farashan Islands 173
Blepharis spiculifolia (Acanthaceae), thorns on Socotra 170
Boehmeria nivea (Urticaceae; ramie), fibre cells 69
Bose, Jagadis Chandra 45
Boswellia spp. (Burseraceae)
  resin (frankincense) 71
  on Socotra 165, 172, 174–6 (*Plate 16*)
Bouteloua gracilis (blue grama), subject of mimicry 42
bracken dienone
  mutagenic 111–13
  see also Pteridium aquilinum
Bradysia impatiens (Diptera; fungus gnat) 125
Brassica napus (Brassicaceae; oilseed rape), idioblast engineering 70
Brassicaceae, glucosinolates 84, *85*
Bremisia tabaci (silverleaf whiteflies), suppression of jasmonate activity 138
browsers, difference from grazers 25–6
browsing, and recovery growth 24
Brycon petrosus (characid fish), herbivory 12
bundle sheath, L2-derived cells 20
Byttneria aculeata (Sterculiaceae), colonization mimicry 39

C
caecotrophy 15
Caesalpinia schlechteri (Caesalpinaceae) 60
Caladium (Araceae), leaf markings 39, 67
Calamus sp. (rattan palms) thorns 56, *57*
calcium salts, use in defence mechanisms 65–7, 121
calotropin 107, *108*
Calotropis procera (Apocynaceae; apple of Sodom), latex 106, *108*
calystegines 86
camphene *106*
Cannabis sativa (Cannabinaceae; hemp) 51, 65
Capparis spp. (Capparaceae) 57, 60, 173
Capreolus capreolus (roe deer), colour-selective browsing 32
Capsicum spp. (Solanaceae; chili), capsaicin 81, *82*
Caralluma spp. (Apocynaceae), crypsis 41
carbohydrates, nutritional accessibility 21
carbon, nutritional availability 1, 21
carbon cycle 2–4, 20
carbonyl groups, α,β-unsaturated 102, *103*
cardenolides (cardiac glycosides) 91, 108–9
Cardinalis spp. (cardinals), eating chilis 81
Carissa ovata (Apocynaceae), thorns on New Caledonia 60
carnivores *see* predators
Carpinus betulus (Betulaceae; hornbeam), parasitized leaf 9 (*Plate 1*)
Casuarina spp. (Casuarinaceae), infected by mistletoe 42, *43*
catnip syndrome 104
Catalpa spp. (Bignoniaceae), extrafloral nectaries 143
Cecropia (Cecropiaceae), food bodies 144
Cedrus deodara (Pinaceae; Himalayan cedar), resin 105
cellulose, nutritional accessibility 21
Central and South America
  enduring plant defences 160
  reptile folivores 12
  spiny palms 58
  three-toed sloth 162
Cerastium glomeratum (Caryophyllaceae; mouse-ear chickweed), sticky trichomes 50
Chazeau, Jean 77
chemical defences *see* defence chemicals
chemical polymorphism, in direct defence 142–3
Chionochloa macra, mimicry by spear grass seedlings 159
chlorophyll 21, 29–30
chloroplasts, migration/alignment 29
Chrysanthemum (Asteraceae), pyrethrins 104
chymotrypsin inhibitors 118
cichlid fish 12
Cirsium spp. (Asteracea; thistles), prickles 58–9
Cissus spp. (Vitaceae), tongue irritants 167
citral 104
cliffs, as refugia 175–6 (*Plate 16*)
Clostridium perfringens, germ warfare 62
Cnemidophorus spp. (whiptail lizards), herbivory 12

Cnidoscolus spp. (Euphorbiaceae; wild chayas), sting variability 52–3
Cnidoscolus urens (Euphorbiaceae; mala mujer), sting 54
co-evolution, of plants and herbivores 26–7, 56, 61, 75–6, 89, 101, 144, 156, 178
cocaine 86
codeine 88
Codiaeum variegatum (Euphorbiaceae; croton), orange leaf surfaces 30
Coffea spp. (Rubiaceae; coffee), cavity domatia 149
COI1 protein, jasmonate receptor 127–8
collagens, tannin binding 100
Collembola (springtails), as herbivores 10
colleters 49
colonization mimicry 38
colour vision 16, 31–2
Commelinaceae (spiderworts), silica deposition 64
Commiphora spp. (Burseraceae)
  in Arabia and on Socotra *167* (*Plate 13*), 168, 170, 173–4
  resin (myrrh) 71
compensatory (recovery) growth 24
conifers *see* gymnosperms; *Pinus*
Conium maculatum (Apiaceae; hemlock), warning pattern 33
Conolophus spp. (Galapagos land iguanas) 12
contraceptives 111
coppicing 24
Corokia cotoneaster (Cornaceae), divarication 158
coronatine, molecular mimic of JA-Ile 127
corontatine insensitive 1 coi1 gene 127
Cotesia marginiventris (wasp), parasitoids 150, 152
Croton socotranus (Euphorbiaceae), browsed by goats 167
crustaceans, would-be folivores 137
crypsis, strategies 29, 40–1, 44
Cryptolepis intricata (Apocynaceae), stunting on Socotra 168
crystals, in defence mechanisms 65–8
Ctenopharyngodon idella (grass carp), herbivory 12
Cucurbitaceae (cucumber family), silica deposition 64
cuticle, role in defence 47
Cyanea (Campanulaceae), on Hawaii 160
cyanide 83–4
Cyanocitta spp. (jays), prey polymorphism 158
cyanogenic glycosides *82*, 83, *84*, 111
Cycas (Cycadaceae), cycasterone 110
cycasterone 110
Cymbopogon citratus (Poaceae; West Indian lemon grass), monoterpenes 104
Cynomys ludovicianus (black-tailed prairie dog) 79
Cyperaceae (sedges), silica deposition 64
cyprinid fish (*le capitaine*), herbivory 12
Cypripedium spp. (orchids), leaf spotting 34
cytochrome *c* oxidase, inhibited by cyanide 83, *84*
cytochrome P450 enzymes, in detoxification 100
Czenspinskia transversostriata (mite) 149

D
Dactylorchis spp. (orchids), leaf flecking 34, *35* (*Plate 4*)
damage
  affects mRNA levels 128
  mechanical vs insect damage 7–8, 120–1
  phased response 130
  tolerance 24–6, 177
damage-associated molecular patterns (DAMPS) 130
damage-induced responses, and growth 133–4
danthonioid grasses (Poaceae; pampas grasses), leaf abscission 26
defence chemicals
  abundant in plant kingdom 101
  alternative roles 80
  cost to produce 80
  detoxification 75, 100–3, 178
  and herbivore physiology 81–3, *87*, 178
  importance 11
  indestructibility 115
  inducibility 120, 124
  latitude clines 90–2
  selection pressures 75
  storage in photosynthetic cells 20
  as trophic barriers to insects 75
defence strategies
  anachronistic 155–61
  common syndromes 59
  in endemic Socotran flora 166–72
  energetics 178–9
  environmental factors 58–9
  indirect 142
  inducibility *119*, 120, 124, 179
  reaction time 115
  time factors 178
  varied and highly adapted 179–80
DELLA proteins, growth control *134*
Dendrocnide moroides (Gympie nettle), sting 52, *53* (*Plate 6*)
Dendrophthoe homoplastica (mistletoe), host mimicry 43
Dendrosicyos socotrana (Cucurbitaceae; cucumber tree), on Socotra 166
Deschampsia caespitosa (tufted hair-grass), silica deposition 64
detoxification 100–3
detritivory, vs herbivory 136–8
dhurrin 83, *84*
Dichrostachys dehiscens (Fabaceae), on Socotra 169–70
Dicoma cana, on Socotra 171

Dicrostonyx groenlandicus (collared lemmings), population dynamics 141
Dieffenbachia, raphides 70
digestive tract
co-evolution with leaves 27, 89, 178
detoxification 115
effect of alkaloids 89
effect of opiates 88
of grazers 25
symbionts 15
target of leaf defences 15
Digitalis lanata (Plantaginaceae; Balkan foxglove) 108
digoxin 108–9
Dionaea muscipula (Droseraceae; Venus flytrap) 92
Diplura (bristletails) 10
Dipsosaurus dorsalis (desert iguana), folivory 12
Discaria toumatou (Rhamnaceae), thorns in New Zealand 157
Distichodus rostratus ('grass eater' fish) 13
diuretics, as defence compounds 22
divarication 156–7, 160
domatia 143–9
domestic livestock
diurnal feeding 34
locoism 87
on Socotra 166–8
Dracaena cinnabari (Dracaenaceae; dragon's blood resin) 165
Dracaena ombret (Dracaenaceae), cliff refugia 175
Dumontet, Vincent 90

E
ecdysones *108*, 110
Echinochloa spp., fish herbivory 13
Echium wildpretii (Boraginaceae; Mount Teide bugloss), cotyledon markings 40
Eclipta (Asteraceae), fish herbivory 12
Ehrlich, P.R. 75–6
Eichhornia crassipes (Pontederiaceae; water hyacinth), calcium oxalate crystals 66
electrical signals, in plant defence 132
enzyme inhibition, blocking digestion 86–7, 117, 119
epidermis, role in defence 48
epiothiospecifier (ESP) proteins *85*
Equisetum spp. (Equisetaceae; horsetails), defences 48, *49*, 64, 115, 135–6
Erica carnea (Mediterranean heather), phenolics 99
Erigeron (Asteraceae), fish herbivory 12
Eriosyce chilensis (cactus), leaves reduced to thorns *57*
Erythroxylum coca (Erythroxylaceae; coca), alkaloid deprotonation 86
Ethiopia
diurnal feeding by domestic livestock 34
geladas 64, *65*
warning signals *35* (*Plate* 3), *37*
eucalyptol *82*
Eucalyptus spp. (Myrtaceae)
cyanide-producing species 83
folivory 100, 162
kino (gum) 71
marsupial detoxification mechanisms 100
parasitized by mistletoe 43
Euphorbia spp. (Euphorbiaceae)
E. arbuscula
caustic latex 107, *109*
on Socotra *167* (*Plate 13*)
E. cactus, white warning patterns 36, *37*
E. collenetteae 173
E. cyparissias (cypress spurge), latex 107, *110*
E. poisonii, resiniferatoxin 107
E. spiralis, on Socotra 167, *167* (*Plates 13* & *15*), 171
evergreens, winter food source 13, 89
exposure colouration 31–2
extrafloral nectaries 124, 143–7
exudates, role in defence 70–2

F
Fagus sylvatica (Fagaceae; beech), domatia 30, 149
Festuca novae-zelandiae (Poaceace), mimicry 42, 159
fibre cells (scelerenchyma), role in defence 68–9
Ficus spp. (Moraceae; figs)
epidermal lithocysts 65
latex production 72
leaf indentations 41
fish, herbivorous 12–13
flavonol glycosides, in Venus flytrap 92
Flindersia dissosperma (Rutaceae; scrub leopardwood), relict defences 160
Foley, William 101
folivores *see* herbivores; invertebrate herbivores; vertebrate folivores/herbivores
folivory
in amphibians 11
in birds 13, 173
in fish 12–13
in mammals 14–17, 55, 64–5, 162
in marsupials 16, 61, 64, 72, 101, 162
in reptiles 11
see also herbivory
food bodies 144
foods, calorific content vs digestibility 22
foraging, year-round 13
forest trees
growth strategy 133
release from vertebrate pressure 162
frankincense 71
see also Boswellia spp. (Burseraceae)

G

Gabon
  ant plant 147
  fish herbivory 12–13
gall-forming insects 8–9, 134
gallic acid *98*, 102
gastrointestinal tract *see* digestive tract
gastropods *see* molluscs
Gazella gazella subsp. *farashani* (Farasan gazelle) 166, 173
gazelline antelopes, and acacia thorns 61
gelada, phytoliths in faeces 64, *65*
Geranium viscosissimum (Geraniaceae; sticky geranium) 49
Glander, K.E. 87–8
Glenny, David 41–2
Gliricidia sepium (Fabaceae; madera negra), defence chemicals 88
glochids 55–6
glucosinolate sulphatase 86
glucosinolates
  in *Arabidopsis* 70
  chemistry 84–6
  jasmonate inducible 124
  sulphur-rich defence compounds 48
  typical of Brassicaceae 84, *85*
  and woodlouse detrivory 137–8
glue, as defence mechanism 49
glutathione, depletion by electrophiles 102
goats, on Socotra *119*, 166–7
Gopherus spp. *see* tortoises
grasses
  apical meristems protected 26
  defence strategies 180
  mimicry by other plants 41, 159
  silica deposition 64
  tolerance to damage/grazing 25–6
grayanotoxins, from rhododendron *87*, 89
grazers, digestive tract 25
green leaf volatiles (GLVs) *see* volatiles
Greenwood, R.M. 156–7
growth
  effects of light 133
  and jasmonate 132–4
  recovery from browser feeding 24
  trade-off with defence 133–4
  wound-induced responses 133–4
guard cells 48
gum
  jasmonate-inducible production 124
  role in defence 71
Gunnera (Gunneraceae), slug/snail feeding 10
gymnosperms, defence strategies *31*, 32, 105–6, 124, 180

H

Hakea (Proteaceae), sclereids 69
Hawaiian archipelago, anachronistic defences 5, 159–61
Helichrysum depressum (Asteraceae), 'playing dead' 158
Heliconius spp. (heliconid butterflies), deceived by leaves 38
Helix spp. (garden snail) 10, 54
Helleborus spp. (Ranunculaceae; hellebores)
  leaf layers L1—L3 *19*
  protoanemonin 111
Heracleum mantegazzianum (Apiaceae; giant hogweed) 95–6
herbivores
  co-evolution with plants 26–7, 56, 61, 75–6, 89, 101, 144, 156, 178
  extinction and introduction 155
  feeding time and warning colouration 33
  highly specialized 1
  importance of teeth and mandibles 62
  night vision and nocturnal activity 33
  tannin-binding proteins 100
  types 4–5, *136*, 137
herbivory
  and the carbon cycle 2–4
  escape from 177
  induced responses 120
heterophylly, on Socotra 171–2
Hevea brasiliensis (Euphorbiaceae; rubber tree), latex 107
hexenals 75, 150–1
Hibiscus noli-tangere (Malvaceae), irritant hairs *168 Plate 14*, 169
histamine, in plant stings 54
holdfast, cliff trees 175–6
hormone mimics 110–11
humans
  cooking of plant foods 181
  and ecosystem change 141–2
  farming activities 3, 180–1
  not obligate leaf eaters 1
Humulus lupulus (Cannabinaceae; hop), calcium carbonate crystals 65
Hyacinthoides non-scripta (Hyacinthaceae; bluebells), alkaloids 86
hydathodes 19–20, 48
20-hydroxyecdysone, insect moulting *108*, 110
hydroxyflavan-3-ol *98*
Hymenophyllum spp. (Hymenophyllaceae; filmy ferns) 17–19
Hyoscyamus niger (Solanaceae; henbane), hyoscyamine *87*, 89
Hypericum perforatum (Hypericaceae; St John's wort), hypericin 96

I
idioblasts, role in defence 67, 70, 178
Iguana iguana (green iguana) 12
Ilex aquifolium (Aquifoliaceae; holly)
  anachronistic prickle gradient 155
  theophylline 89
Impatiens (Balsaminiferae), fish herbivory 12
Indigofera argentea (Fabaceae), sand armour 63 (*Plate* 7)
insectivores *see* predators
insects
  colour vision 32
  damage to mandibules xx
  egg-laying 138–9
  generalists vs specialists 11
  herbivory 3–4, 7
  mandible wear 67
invertebrate herbivores
  effect of leaf surface topography 47
  feeding strategy 129
  mandibles 63–4, 67, 71, 73, 107, *108*, 121
  relative importance 3–4
  on Socotra 172
  types 5–11
ion channel receptors 81–2
iron, chelation by tannins 100
islands
  faunal history 90–1
  herbivore populations 165–6
isoprene 104
isothiocyanate 84–6

J
Janzen, Daniel 23, 76, 120, 146–7, 160
Jardin du Talèfre, vertebrate fauna 161–2
jasmonate
  biological roles 120, 122, 124–6
  conjugates 127
  and defence inducibility 179
  evolution 135–8
  gene activation 128
  and growth 132–4
  mutant studies 124–6
  speed of synthesis 126-7
  structure and synthesis 122, *123*
jasmonate signalling
  in the attacked leaf 126–7
  couples direct and indirect defences 153
  electrical signals 132
  origin *136*
  pathway 128–30
  propagation to distal tissues 129–31
  relevance to agriculture 181
  suppression by herbivores 138–9
  velocity of rapid wound signal 131–2
Jatropha unicostata (Euphorbiaceae), on Socotra 167
JAZ proteins (jasmonate responses) 128, *129*, *134*
Juglans spp. (Juglandaceae; walnut) 93
juglone 92, 93, 101
Justicia rigida (Acanthaceae), spininess 170
juvenile forms, need for defences 31–2, 163

K
Kalanchoe marmorata (Crassulaceae), red warning
    pattern 34, *35* (*Plate* 3)
Kingdon, J. 166, 168
kino, produced by *Eucalyptus* (Myrtaceae) species 71
koalas, eucalyptus browsing 100
kudus, and acacias 154

L
Laportea spp. (Uritcaceae; nettles) 55
Lasius niger (black garden ant) 142–3
latex-producing plants, white warning patterns 36
latexes
  caustic 107, *109–10*, *167* (*Plate* 13)
  chemistry 71–2
  colour 107
  nickel-containing 77–8, 107
  terpene macropolymers 106–7
laticifers 71–2, 124
leaf cell types 6, 18–21
leaf colouration 29–33, 37–40
leaf defences
  definition 23–4
  evolution 17
  protect protein and amino acids 22
  seasonal 13
  selection pressures 11, 23–4, 32–3
  social context 23–4
  specificity 11
  target digestive tract 15
leaf economics spectrum 18
leaf miners 8–9, 39
leaf patterns/markings
  artificial 32, 39
  association with strong defences 66–8
  as camouflage 36–7
  types 30
leaf surfaces
  prickly 55
  slippery 47
  sticky 49
leaf-chewing insects, type of damage inflicted 9
leaf-cutting ants, mandibles reinforced with zinc 65
leaves
  attractiveness of new foliage 22, 31–2
  chemical capabilities 177
  evolutionary history 17–18

as food source 1, 21–3
layers L1-L3 19–20
lifespan 18
loose definition 17
mechanical wounding vs insect damage 120–1
multifarious nature and diversity 18
not primary storage organs 177
nutritional attractiveness to herbivores 177
patterned *see* leaf patterns/markings
polymorphism 32–3, 158, 171–2
protein content 21, 181
relief from vertebrate pressures 162–3
surface:volume ratio 17
tolerance to damage 25–6, 177
vascular systems 20, 130
Leptadenia pyrotechnica (Apocynaceae), fibre cells 69
Leptinotarsa decemlineata (Colorado beetles) 118
Leucas spiculifera (Lamiaceae), on Socotra 170
Levin, D.A. 90, *91*
levopimaric acid, in resins *106*
light, effects on growth and defence 133
lignin 21–2, 68
Ligustrum obtusifolium (privet), reactive phenolics 96
Limax cinereoniger (banana slug) 10
linalool, engineered into *Arabidopsis* 104
lithocysts 65
Loasaceae, trichomes 49
locoism, in livestock 87
Lotus corniculatus (Fabaceae; birdsfoot trefoil) 100
lox jasmonate mutants *125*
Ludwigia (Onagraceae), fish herbivory 12
Lycium sokotranum (Solanaceae), on Socotra 168–9

M
Macaranga spp. (Euphorbiaceae)
Beccarian food bodies 144
coloured latex 107
symbiotic ants 47
Machlura cochinchinensis (Moraceae), thorns on New Caledonia 60
Macropteranthes keckwickii (Combretaceae; bullwaddy) 160
Madia elegans (Asteraceae; tarweed), insect trapping 50
mammals
folivory 14–17, 55, 64–5, 162
role in seed dispersal 172
see also domestic livestock
Manduca sexta (tobacco hornworm) 85, 124–5
Marantochloa spp. (Marantaceae), nyctinastic movements 44
Marsilea crenata (Marsileaceae), silica content 64–5
marsupials
Eucalyptus detoxification mechanisms 100–1
as folivores/herbivores 15–16, 43, 162
Maytenus sp. (Celastraceae) 173
Mazama americana temama (brocket deer) 146–7
Medicago truncatula (Fabaceae; barrel medic), calcium oxalate crystals 67
Melicytus alpinus (Violaceae) 157
Mentha spp. (Lamiaceae; mints) 48, 104
menthol 104
metals, accumulation by plants 77–9
8-methoxypsoralen, photoactivated toxin 95, *96*
methyl jasmonate *see* jasmonate
Metopium toxiferum (poisonwood tree) 94
Metynnis spp. (silver dollar fish), herbivory 12
Microtis agrestis (field voles) 64, 162
Miller, A.G. 174
mimicry
by mistletoe 42–3
of grasses and other plants 41–2
of butterfly eggs 138
selective advantage 43
of soil, sticks, and stones 40–1
Mimosa pudica (Fabaceae), leaf movements 44–5
mistletoe (Loranthaceae), host mimicry 42–3
mites (Acari)
detoxification of plant chemicals 10
domatia 147–9
moa-nalos, anachronistic defences 159
moas 26, 42, 55, 155–7
molecular mimicry, enzyme inhibition 86–7
molluscs, herbivorous 10
monkey puzzles, selection for spininess 60–61
monoterpenes 100–1, 104, *105*
morphine 88
Morus australis (Moraceae; Chinese mulberry), calcium crystals 67–8
mountain gorillas, eating of nettles 55
mucilage, role in defence 71
Muehlenbeckia ephedroides (Polygonaceae), 'playing dead' 158
Müllerian bodies (on *Cecropia*) 144
mustard oil bomb 84–5
mutagenesis 111, *112*
MYC transcription factors *128–9*, *134*
myrmecophily 143
myrosinases 70, 85
myrrh 71
see also Commiphora spp. (Burseraceae)
Myzus persicae (green peach aphid) 85

N
Na^+/K^+-ATPase 106, 108–9
inhibitor 107
1,4-naphthoquinones 92, 101
Near Eastern thorny flora 59
Nepeta cataria (Lamiaceae; catnip), nepetalactone 104, *105*

nettles *see Urtica*; Urticaceae
Neuracanthus aculeatus (Acanthaceae), on Socotra 171
New Caledonia
  flora 55
  indigenous fauna 59–60
  plant defence chemistry 90
  trophic islands 76–8
New Zealand
  anachronistic defences 155–61
  divarication 156–7
  extinct herbivores 155–6
  indigenous fauna 26
  leaf form and patterning 157–8
  spear grasses (*Aciphylla* spp., Apiaceae) 158–9, 173
nickel, hyperaccumulation 77
Nicotiana attenuata (Solanaceae; wild tobacco) 89
Nicotiana tabacum (Solanaceae; tobacco), trichomes 49
nicotinamides, as reducing agents 101–2
nicotine 80, *87*, 89
nitrile specifier proteins (NSPs) 86
nitrogen 26, 79, 88–9, 92
norsesquiterpenes 113
noscapine 88
Nothofagus spp. (Nothofagaceae; southern beeches), vulnerable to introduced herbivores 24
nyctinastic movements 44
Nymphaea odorata (Nymphaeaceae; fragrant water lily), sclereids 69

O
Ocotea foetens (Lauraceae), domatia 148
Odocoileus hemionus (mule deer), salivary proteins 100
Olea europaea (Oleaceae; olive), reactive phenolics 69, 96
oleoresins, monoterpene components 105–6
oleuropein 96, *97*, 101
oligogalacturonic acid 122
Onobrychis spp. (Fabaceae) 59
Ononis repens (Fabaceae; rest-harrow), sand armour 63
opiates, role in nature 88
opioid receptors, in the nervous system 88
Opisthocomus hoazin (hoatzin) 14
Opuntia spp. (Cactaceae; prickly pear) 17, 55–6, 176
Othonna herrei (Asteraceae), mimicry 42
ouabain 109
oxalate crystals 66–7, 121
Oxalis acetosella (Oxalidaceae; common wood-sorrel), nyctinastic movements 44
12-oxophytodienoic acid (OPDA) *123*, 135
oxygenation, as detoxification mechanism 100

P
paclitaxel, jasmonate inducible 124
palisade cells 20
palms, defences 56, *57*, 58, 61–2
Pandanus spp. (Pandanaceae), prickly leaf-edges 60
Papaveraceae, alkaloids 88
papaverine 88
Parsonsia heterophylla (Apocynaceae; jasmine), heterophylly in New Zealand 158
Passiflora spp. (Passifloraceae; passion flowers), colonization mimicry 38 (*Plate* 5)
pavement cell 19
pearl bodies 144
Pediocactus papyracanthus (grama grass cactus), grass mimicry 42
pentagalloyl glucose 98
Peperomia prostrata (Piperaceae), calcium oxalate crystals 67
Periplus, account of Socotra 173–4
pesticides 180–1
Petaurus spp. (gliders) 72
petiole feeders 88
Petrogale concinna (Macropodidae; narbalek, little rock wallaby) 64–5
Peucetia spp. (lynx spiders) 49–50
Pflanzen und Schnecken (Stahl, 1888) 10
Phaseolus lunatus (Fabaceae; lima bean), defence chemicals 150
phenolics 90–6, *94*, 98, 99
Philaenus spumarius (froghoppers), effect on plant growth 7–8
phloem 7, 20, 132
Phoenix dactylifera (Aracaceace; date palm), microbial flora 62
Phormium tenax (Agarvaceae; New Zealand hemp), fibre cells 68
phosphorus, use in defence chemicals 79
photoactivation of toxins 95–6
photosynthetic cells as a foodsource 21
phytoanticipins 79–80, 119
phytoecdysones 110–11
phytoliths 64–5
Phytoseiulus persimilis (predatory mite) 150
Picea abies (Pinaceae; Norway spruce), defences *31*, 32
Pieris brassicae (large cabbage butterfly) 139
Pieris rapae (small cabbage butterfly) 11, 86, 121, 127–8
PIF transcription factors *134*
pigments, phenolic 30
pinene *106*
Pinus cembra (Pinaceae; Arolla pine), resin 105
Pinus nigra (Pinaceae; Mediterranean pine), phenolics 99
plant chemicals *see* defence chemicals
plants
  chemical diversity 75
  colonized by invertebrates 5
  see also ant—plant interactions

damage by herbivory 24
plasma membrane depolarization 132
plastids, jasmonate synthesis 132
Pliny the Elder 93
plumbagin 92–3, 101
1,4-polyisopropenes 106–7
polyphenol oxidases 93–4
Polypodium plebeium (Polypodiaceae), ant nectaries 143
Polypodium vulgare (Polypodiaceae), phytoecdysones 110
Porcellio scaber (common woodlouse), folivory vs detritivory *137*
predators
exposed on pale background 31
impact on herbivory 141–54
influenced by plants 142
removal from environment 141
reward mechanisms 24, 180
volatile-driven behaviour 150, 153
primates, folivory 16–17, 22–3, 55
proteinase inhibitors 115, 118, 122–4, 132
protoanemonin 111
Protura (boneheads), lack of herbivory 10
Prunus avium (Rosaceae; wild cherry), extrafloral nectaries 143, *144 Plate 11*
Prunus spinosa (Rosaceae; blackthorn), thorns 56
Pseudocheirus peregrinus (ringtail possums), feeding on mistletoes 43
Pseudochirops archeri (green ringtail possum) 72
Pseudomyrmex spp. (ants), on vachellias 145–7
Pseudopanax (Araliaceae), anachronistic defence features *156*
Pseudopanax crassifolius (Araliaceae;lancewood), leaf markings *156*, 157
Pseudophyllanax imperialis (giant New Caledonian grasshopper) 65
Pseudotsuga menziesii (Pinaceae; Douglas fir), resin 105–6
Pseudowintera colorata (Winteraceae; mountain horopito), mimicry 42
Psychotria spp. (Rubiaceae), on ultramafic soil 77–8
ptaquiloside, bracken toxin 111, *112*, 113
Pteridium aquilinum (Dennstaedtiaceae; bracken fern)
defence chemicals 83, 111–13
extrafloral nectaries 143
toxicity 15
Pulmonaria officinalis (Boraginaceae; common lungwort), leaf spotting 39, *40*
Pupalia lappacea (Amaranthaceae), seed dispersal by mammals 172
pyrethrins 104
pyrrolizidine alkaloids 143
Pyrus communis (Rosaceae; pear), stone cells 69
pyruvate dehydrogenase 115–16

Q
quality reduction mimicry 37–8
Quercus geminata (Fagaceae; sand live oak), leaf shedding 8
Quercus robur (Fagaceae; pedunculate oak)
calcium oxalate crystals 66
mites 148
tannins 97

R
Ranunculus spp. (Ranunculaceae; buttercups), protoanemonin 111
Rauh, W. 58
Raven, P.H. 75–6
recovery growth, and browser feeding 24
reptilian herbivores 12, 32, 41, 56, 173, 176
resiniferatoxin 107
resins 71, 105–6, 124
resource availability hypothesis 133
rhinos, black vs white 26
Rhododendron maximum (Ericaceae; great laurel) 49
Rhytidocaulon spp. (Apocynaceae), mimic dead wood 41
Ribes spp. (Rosaceae; brambles), prickles 56
ribosome inactivating proteins (RIPs), inducible 116
ribulose bisphosphate carboxylase/oxygenase (Rubisco), nutritional value 21, 116
Ricinus communis (Euphorbiaceae; castor bean), slippery cuticles 47
rodents, herbivory 16, 61, 64, 81, 83
Root, Richard 11
Rosa spp. (Rosaceae; roses), prickles 56
rotenone 88
Rumex alpinus (Polygonaceae; alpine dock), warning pattern 34
rumination, importance in herbivory 15
Ryan, C.A. 118, 122

S
salicylate pathway 138
salivary components, role in defence induction 120–2
Salix sp. (willow), regrowth after damage 24
sand, use in defence 62–3
Sanseviera spp. (Asparagaceae), fibre cells 68
saponins 108
Saudi Arabia, plant defences 36, 37, 62–3
Sauromalus obesus (chuckwallas) 12
sclereids 69
scopolamine 89
Sebertia acuminata (Sapotaceae; nickel tree) 77–8 (*Plate 8*), 107

seed dispersal, by mammals 172
Selaginellaceae (spikemosses), colour patterns 29–30
selenium, accumulation 79
Semecarpus spp. (Anacardiaceae; false mahogany), phenolics 94
Senebier, Jean 59
Senecio jacobaea (Asteraceae; ragwort) 142–3
Senecio viscosus (Asteraceae; stinking groundsel) 50
sequestration, as protection mechanism 109
sesquiterpenes 111–13
Setaria verticillata, seed dispersal by mammals 172
sieve tubes, in phloem 20–1
silica, use in defence 62–5, 120–1
Silphium plant, contraceptive properties 111
slugs 10, 129, 172
Smilax spp. (Smilacaceae), spininess 58
Sminthurus viridis (lucerne flea) 10
snails 10–11, 54–5, 172
social context, and investment in defence 23–4, 57–8, 177
Socotra
  cliff trees 174–6
  endemic flora and fauna 165–6
  herbivores 172–4
  human activity and livestock 165–7
sodium, essential nutrient for animals 22
soils, ultramafic 77
Solanaceae, defence strategies 88–9, 180
solanine 86
Solanum berthaultii (Solanaceae; wild potato), aphid trapping 94
Solanum tuberosum (Solanaceae; potato), steroidal alkaloids 86
Solidago altissima (Asteraceae; goldenrods), insect herbivores 5, 7
spininess
  association with climate severity 58
  defence against herbivores 59
  effects of plant society 57–8
  environmental aspects 58
  eye-piercing 158, *159*
  microbial flora 61–2
  multiple roles 56–7
  in New Zealand spear grasses 158–9
  related to Pleistocene megafauna 160
  selection pressures 59, 61
  in Socotran endemic flora 169–70
  visibility 56
Spodoptera exigua (Lepidoptera; beet armyworm) 67, 86, 89, 150–2
Spodoptera littoralis (Egyptian cotton leafworm) 11, 107, *110*, *119* (*Plate 9*), *125* (*Plate 10*), 139
Stahl, E. 54–5
starch, variable levels 21
Stigmella sp. (Lepidoptera; pygmy moth), on hornbeam leaf 9
Stilbosis quadricustatella (leaf miner) 8
stings
  chemistry 53–4
  most powerful 52, *53*, 55
  polymorphism 53
  potential therapeutic use 55
  relatively rare 52
  structure 52
  vertebrate targets 54–5
Stipagrostis (Poaceae), sand as protectant 62
stomata, rarely used as entry point 19, 48
stone mimicry 41
sucking insects 7–8, 48, 69–70, 147
sulphur, use in defence chemicals 79
Swainsona canescens, mannosidase, locoism 87
swainsonine 86, *87*
Sykes, Bill 55
symbionts, in herbivore digestive tract 15, 21, 115

T
tannins 97–100, *98*, 101–2
taste receptors (bitter, TAS2R proteins) 81, *82*, 83
Taxus baccata (yew)
  holdfasts 176
  20-hydroxyecdysone 110
  toxins 89, 124
terpenes
  chemical complexity 104
  as defence chemicals 104–13
  in glandular trichomes 48–9
  mutagenic 111
  polymerization 104
  in resins and latexes 71, 105–6
Tetranychus urticae (two-spotted spider mite) 10, 150
thebaine 88
Theobroma cacoa (cocoa; Malvaceae), colour of new leaves 31
theophylline 89
Theropithecus gelada (gelada), phytoliths in faeces 16, 64, *65*
thiamin (vitamin B1) 115
thiaminases 111, 115
thorn cushion, geographical distribution 59
thorns, spines, and prickles *see* spininess
threonine, targeted by plant proteins 115–16
threonine deaminase 115–16, *117*, 124
thumbergol *106*
Tilia × europaea, (Tiliaceae; common lime), domatia 149
tolerance to damage, critical to plant survival 24–6, 177
tomato
  jasmonate mutants 124–5

proteinase inhibitor production 116, 118, 122–4, 132
threonine deaminase 116, *117*, 124
tortoises
extinct 60, 76
herbivory 12, 56, 173–4, 176
indigenous to Indian Ocean islands 165–6, 174
protection from 172, 174
Toxicodendron spp. (Anacardiaceae; poison ivy, poison oak), phenolics 94–5
toxins, detoxification 75, 100–3, 178
Tragia spp. (Euphorbiaceae), sting 53, 168–9
transient receptor potential (TRP) proteins 81–3
see also TRP receptors
Trichogoniopsis adenantha (Asteraceae), sticky leaf surfaces 49
trichomes
chemical defences 48–9, 94
as insect traps 49–50
jasmonate inducible 124
leaf surface defence cells 20, 47–52
mechanosensitivity 51–2
physical defence 49
trapping of sand grains 63
Trichoplusia ni (cabbage looper) 109
Trichosurus vulpecula (brushtail possums), feeding on mistletoes 43
Trifolium repens (Fabaceae; clover) 32, 100
Trimerotropis spp. (grasshoppers) 90
triterpenes 108–11
tritrophic interactions 142–3, 150
trophic islands 76–8
TRP receptors
TRPA1 (glucosinolate receptor) 81, 85
TRPM8 (terpene receptor) *82*, 83
TRPV1 (capsaicin receptor) 81, *82*, 107
trypsin inhibitors 118
Tumlinson, James 151
Turlings, Ted 151
Tussilago farfara (coltsfoot, Asteraceae), tolerance to damage *25* (*Plate 2*)
Tylecodon wallichii (Crassulaceae), toxic succulent 42
Tyria jacobaeae (cinnabar moth) 142–3

U
ubiquitination, in jasmonate signalling 128
ungulates
and acacias 154, 173
colour vision 32
defence against 61
indigenous, on islands 173
rumination 15
Urera baccifera (cow-itch) 55
Uromastyx ('spiny-tailed' lizards) 12
Urtica dioica (Urticaceae; common nettle)
grazed by snails 54
sting 52–5
trichomes 51
Urtica ferox (Urticaceae; ongaonga nettle), powerful sting *53*, 55 (*Plate 6*), 158
Urtica thunbergiana (nettle), sting chemistry 54
Urticaceae (nettles)
induced sting development 120
silica deposition 64
urushiols 94–5, 102

V
Vachellia spp. (formerly *Acacia*; Fabaceae)
gums 71
V. collinsii, ant plant 144–6
V. drepanolobium (whistling thorn), ant plant 36, 147
V. ehrenbergiana 173
V. pennivenia, endemic to Socotra 170
white thorns 36, *37*
Varicorhinus cyprinid fish, herbivory 12
VEGETATIVE STORAGE PROTEIN 2 (*VSP2*) gene *119* (*Plate 9*)
VEGETATIVE STORAGE PROTEIN 2 (VSP2) protein, jasmonate inducible 124
vegetative tissues, as bait 120
Venezuela, ecosystem meltdown 141–2
Venus flytrap, putative narcotic 93
vertebrate exudivores 72
vertebrate folivores/herbivores
arboreal 162–3
can reach poorly accessible plants 161, 165
co-evolution 59, 90
endothermic 13–17
escape from 161–3
eyesight and colour vision 32
relative importance 3–4
on Socotra 172–4
types 11–17
Viburnum tinus (Caprifoliaceae; laurustinus), mite domatia 148 (*Plate 12*)
Victoria amazonica (Nymphaceae; Amazonian water lily), underwater defences 58
Viscum album (Viscaceae; mistletoe), does not mimic hosts 43
vitamins, destruction in digestive tract 115
Vitis vinifera (grapevine), domatia 149
volatiles
damage-induced release 75
egg-laying induced release 139
herbivore-induced release 152–3
predator and parasitoid attraction 150–3
volicitin 152

W
Walker-Simmons, Mary-Kay 122
warning patterns 33–4, 36–7
wound response domains *131*

X
Xanthorrhoea spp. (Xanthorrhoeaceae; grass trees), grass mimicry 41
Ximenia americana (Olacaceae), thorns on New Caledonia 60
xylem 7–8, 20

Z
Zea mays (maize), and beet armyworm caterpillar 150
Zenk, Meihard 124